Fundamentals of Communication Theory

Unnikrishna Pillai · Anuj Patel

Fundamentals of Communication Theory

Analog to Digital

Springer

Unnikrishna Pillai
Department of Electrical
and Computer Engineering
New York University
New York, NY, USA

Anuj Patel
Brooklyn, NY, USA

ISBN 978-3-032-20614-5 ISBN 978-3-032-20615-2 (eBook)
https://doi.org/10.1007/978-3-032-20615-2

This Springer imprint is published by the registered company Springer Nature Switzerland AG
The registered company address is: Gewerbestrasse 11, 6330 Cham, Switzerland

Dedicated to the memory of Prof. Dante Youla

Preface

These lecture slides are designed for a semester long undergraduate course in "Fundamentals of Communication Theory". The material presented herein aims to provide a comprehensive introduction to the core principles and mathematical tools essential for understanding modern communication systems.

These slides begin with the foundational concepts of signal, starting with Fourier Series followed by Fourier Transforms, exploring the representation of periodic signals, then extending these ideas to non-periodic signals through Fourier Transforms and ending up with Nyquist's Sampling theorem. The mathematical framework here is important for analyzing signals in both time and frequency domains. Building upon this foundation, these lecture notes transition to the core of communication theory: modulation and demodulation techniques, both in analog and digital domain in presence of noise.

Preface

Next, Amplitude Modulation (AM) is explored in its various forms - Double Sideband Modualtion (DSB), Single Sideband Modulation (SSB), and Vestigial Sideband Modulation (VSB), followed by a detailed study of Angle Modulation, both Frequency Modulation (FM) and Phase Modulation (PM). This is followed by noise analysis in both AM and FM/PM, where advantage of FM/PM over AM is highlighted.

Finally, the course material culminates with an introduction to Digital Modulation Techniques covering topics such as sampling, quantization (Lloyd-Max and uniform), elements of coding and modulation theory which are the building blocks of modern digital communication systems. Finally the classic problem of detection of signal in noise is analyzed in the Binary Pulse Code Modulation (PCM) signaling case, using the minimization of the probability of error criterion for optimum waveform selection in the binary case. This is followed by an analysis of the M-ary

Preface

orthogonal signaling and the corresponding probability of error. PCM analysis shows that the probability of error can be made arbitrarily small through a combination of coding/modulation and as a result error-free transmission is possible to any desired degree of accuracy.

Throughout these notes, theoretical concepts are supplemented with several examples and explanations to foster a deeper understanding. It is our hope that these notes will serve as a valuable resource for students as they navigate through principles of modern communication theory as well as for professors aiming to teach this course supplemented by a set of lecture notes.

Unnikrishna Pillai
Anuj Patel

New York University
May 2025

About the Authors

Unnikrishna Pillai (up240@nyu.edu) received his Ph.D. degree in systems engineering from the University of Pennsylvania in 1985 and MTech in Electrical Engineering from the Indian Institute of Technology, Kanpur, India in 1982. He joined Brooklyn Polytechnic, New York, in 1985 as an assistant professor in electrical engineering, and since 1995 he has been a professor of electrical and computer engineering (currently New York University).
He is the (co-)author of several widely used textbooks including Probability, Random Variables and Stochastic Processes (A. Papoulis and S. U. Pillai) (2002), has over 250 online video lectures, and has published extensively. He is the 4-th most cited Google scholar under 'Radar'.
His research interests include synthetic aperture radar (SAR) imaging, waveform design, power control for wireless mobiles, electronic signal identification, rational and stable approximation, array signal processing, portfolio optimization, and applications of non-linear differential equations for modeling physical systems with time-varying characteristics. He is a Fellow of the IEEE.

About the Authors

Anuj Patel (amp10162@nyu.edu) received his MS degree in Electrical and Computer Engineering from New York University Tandon School of Engineering in 2025 and his undergraduate degree from Vellore Institute of Technology, India (2023). Towards his MS degree, he has worked as a researcher at NYU WIRELESS on Millimeter-Wave Beamforming and MIMO Channel Estimation.
Beyond academics, Patel was also involved with the NYU Tandon Graduate Student Council and the NYU Student Government Assembly. He is the recipient of Emerging Leaders Award, Distinguished Engagement Award and the Tandon Community Excellence Award.

Contents

1 **Fourier Series** 1

2 **Fourier Transforms** 33

3 **Amplitude Modulation** 83

4 **Angle Modulation** 105

5 **Digital Modulation Techniques** 139

1 Fourier Series

Abstract

This chapter provides the mathematical framework necessary for analyzing periodic signals in both time and frequency domains. It begins with the foundational concepts that start with the Fourier Series representation of periodic signals. The focus here is on the representation of periodic signals using Fourier coefficients and the underlying principles such as Fourier's Theorem, Parseval's Theorem, and the Fourier Series expansion of real periodic signals. Along with theoretical development, key concepts with examples are illustrated, highlighting the importance of Gibb's Phenomenon and real-world applications such as vibrato in music and Kepler's equation.

U. Pillai and A. Patel, *Fundamentals of Communication Theory*,
https://doi.org/10.1007/978-3-032-20615-2_1

Table of Contents

1. Fourier Series
1.1 Even and Odd Periodic Signals
1.2 Fourier's Theorem
1.3 Periodic Signals and Fourier Coefficients
1.4 Parseval's Theorem
1.5 Fourier Series Expansion of Real Periodic Signals
1.6 Examples
1.7 Gibb's Phenomenon
1.8 Vibrato in Music
1.9 Reversible Periodic Functions & Kepler's Equation
2. Fourier Transforms
3. Amplitude Modulation
4. Angle Modulation

Fourier Series Fourier Transforms Amplitude Modulation Angle Modulation Digital Modulation Techniques

Fourier Series (Periodic Signals)

A signal $f(t)$ is said to be periodic if it satisfies the condition:

$$f(t) = f(t + kT), \text{for all t} \tag{1.1}$$

where k is any integer, i.e., the signal $f(t)$ repeats itself after every T seconds starting at any t.
The total information of a periodic signal is contained in any $(t, t + T)$.

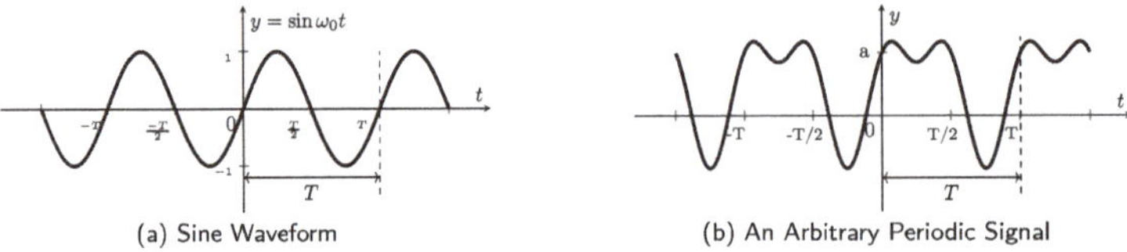

(a) Sine Waveform (b) An Arbitrary Periodic Signal

Figure 1.1: Examples of Periodic Signals

Fundamentals of Communication Theory U. Pillai & A. Patel

Even and Odd Periodic Signals

Even Periodic Signal;

$$f(-t) = f(t)$$

Odd Periodic Signal;

$$f(-t) = -f(t)$$

$\cos\omega_0 t$ and $\sin\omega_0 t$ are even and odd periodic signals respectively for any ω_0.

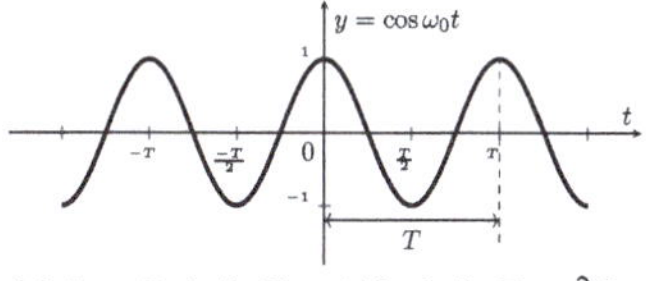

(a) Even Periodic Signal (Cosine), $T = \frac{2\pi}{\omega_0}$

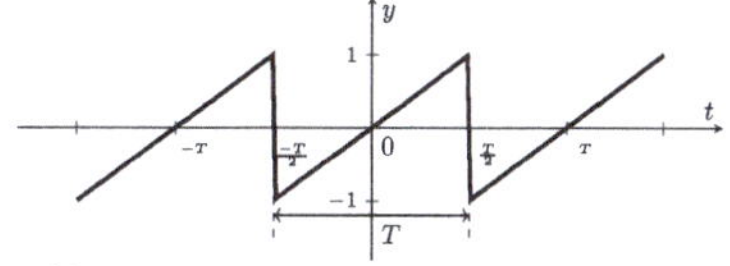

(b) Odd Periodic Signal (Sawtooth waveform)

Figure 1.2: Even and Odd Periodic Signals

Fourier Series Fourier Transforms Amplitude Modulation Angle Modulation Digital Modulation Techniques

Fourier's Theorem

Any periodic signal $f(t)$ with period T has a complex sinusoidal (in terms of sine and cosine terms) representation;

$$f(t) = f(t + kT) = \sum_{k=-\infty}^{\infty} c_k e^{jk\omega_0 t} \tag{1.2}$$

where $\omega_0 = \frac{2\pi}{T}$ is known as the fundamental frequency.

Note: $e^{j\omega_0 t} = \cos k\omega_0 t + j \sin k\omega_0 t$ represents a discrete complex sinusoidal waveform of frequency $k\omega_0$.

Fundamentals of Communication Theory U. Pillai & A. Patel

Fourier Coefficients

For any periodic signal with period T, we have

$$f(t) = \sum_{k=-\infty}^{\infty} c_k e^{jk\omega_0 t}, \ \omega_0 = \frac{2\pi}{T} \tag{1.3}$$

Substituting $f(t)$ into, we obtain

$$\begin{aligned} \frac{1}{T}\int_0^T f(t)e^{-jn\omega_0 t} &= \frac{1}{T}\int_0^T \left(\sum_{k=-\infty}^{\infty} c_k e^{jk\omega_0 t}\right) e^{-jn\omega_0 t} dt \\ &= \sum_{k=-\infty}^{\infty} c_k \cdot \frac{1}{T}\int_0^T e^{j(k-n)\omega_0 t} dt = c_n \end{aligned} \tag{1.4}$$

Note that, (1.4) follows since

$$\frac{1}{T}\int_0^T e^{j(k-n)\omega_0 t} dt = \frac{1}{T} \cdot \frac{e^{j(k-n)\omega_0 t}}{j(k-n)\omega_0}\Bigg|_0^T = \frac{1}{T}\left(\frac{e^{j(k-n)\omega_0 T}-1}{j(k-n)\omega_0}\right) = \begin{cases} 0, k \neq n \\ 1, k = n \end{cases} \tag{1.5}$$

Periodic Signals and Fourier Coefficients

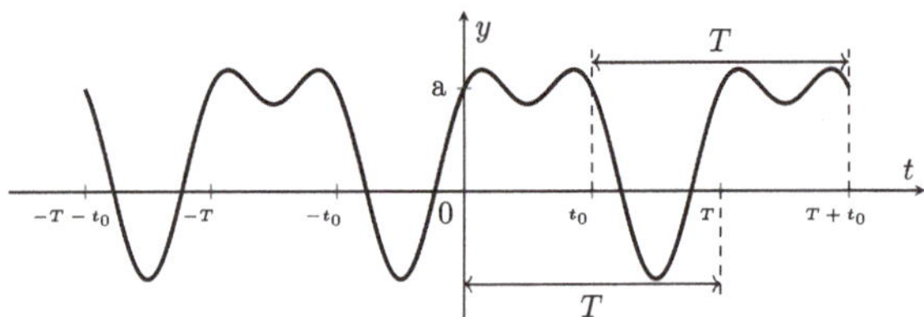

Figure 1.3: Periodic Signal Representation

If $f(t) = f(t+T)$ for all t, i.e., if $f(t)$ is periodic, then

$$f(t) = \sum_{k=-\infty}^{\infty} c_k e^{jk\omega_0 t} \quad \text{where,} \tag{1.6}$$

$$c_k = \frac{1}{T}\int_0^T f(t)e^{-jk\omega_0 t}\, dt \tag{1.7}$$

Periodic Signals and Fourier Coefficients

c_k's can be computed over any duration T. To see this, let

$$\begin{aligned} d_k &= \frac{1}{T}\int_{t_0}^{t_0+T} f(t)e^{-jk\omega_0 t}\,dt = \frac{1}{T}\int_{t_0}^{t_0+T}\left(\sum_{n=-\infty}^{\infty} c_n e^{jn\omega_0 t}\right)e^{-jk\omega_0 t}\,dt \\ &= \sum_{n=-\infty}^{\infty} c_n \cdot \frac{1}{T}\int_{t_0}^{t_0+T} e^{j(n-k)\omega_0 t}\,dt \end{aligned} \tag{1.8}$$

Replace variable t with $t_0 + x$ and using (1.5)

$$\begin{aligned} d_k &= \sum_{n=-\infty}^{\infty} c_n \cdot \frac{1}{T}\int_{0}^{T} e^{j(n-k)\omega_0(t_0+x)}\,dx \\ &= \sum_{n=-\infty}^{\infty} c_n e^{j(n-k)\omega_0 t_0} \cdot \frac{1}{T}\int_{0}^{T} e^{j(n-k)\omega_0 x}\,dx = c_k \end{aligned} \tag{1.9}$$

Periodic Signals and Fourier Coefficients

Thus,

$$d_k = c_k$$

The constant c_k in (1.6) is the amplitude corresponding to the frequency $k\omega_0$.

Here, ω_0 represents the fundamental frequency and $k\omega_0$, $k = \pm1, \pm2, \pm3, \ldots$ represent the overtones. The corresponding amplitudes are denoted by $c_{-1}, c_0, c_1, c_2, c_3, \ldots, c_k, \ldots$ respectively.

Do these amplitudes c_k decrease as k increases?
To answer this question, we can look at the total power contained in all the frequency components in (1.6).

Parseval's Theorem

Consider two periodic signals $f_1(t)$ and $f_2(t)$ with the same period T. Let,

$$f_1(t) = \sum_{k=-\infty}^{\infty} \alpha_k e^{jk\omega_0 t} \quad \text{and} \quad f_2(t) = \sum_{n=-\infty}^{\infty} \beta_n e^{jn\omega_0 t} \tag{1.10}$$

with $\omega_0 = \frac{2\pi}{T}$. From (1.6),

$$\alpha_k = \frac{1}{T}\int_0^T f_1(t)e^{-jk\omega_0 t}\,dt \quad \text{and} \quad \beta_k = \frac{1}{T}\int_0^T f_2(t)e^{-jk\omega_0 t}\,dt \tag{1.11}$$

Combined integral over any duration T gives

$$\frac{1}{T}\int_0^T f_1(t)f_2^*(t)\,dt = \frac{1}{T}\int_0^T \left(\sum_{k=-\infty}^{\infty} \alpha_k e^{jk\omega_0 t}\right)\left(\sum_{n=-\infty}^{\infty} \beta_n^* e^{-jn\omega_0 t}\right) dt \tag{1.12}$$

Fourier Series Fourier Transforms Amplitude Modulation Angle Modulation Digital Modulation Techniques

Parseval's Theorem

Using (1.5),

$$\frac{1}{T}\int_0^T f_1(t)f_2^*(t)\,dt = \sum_{k=-\infty}^{\infty}\sum_{n=-\infty}^{\infty} \alpha_k\beta_n^* \cdot \frac{1}{T}\int_0^T e^{j(k-n)\omega_0 t}\,dt = \sum_{k=-\infty}^{\infty} \alpha_k\beta_k^* \tag{1.13}$$

gives the **Parseval's Formula**. When $f_1(t) = f_2(t)$ in (1.13), reduces to

$$P = \frac{1}{T}\int_0^T |f_1(t)|^2\,dt = \sum_{k=-\infty}^{\infty} |\alpha_k|^2, \tag{1.14}$$

which shows that the total power P of the signal is redistributed over the discrete frequencies $\pm k\omega_0$ for a periodic signal, with $|\alpha_k|^2$ representing the signal power at frequency $k\omega_0$. If $P < \infty$, then $\alpha_k \to 0$ as $k \to 0$. Thus, for any periodic signal, the power content of the overtones eventually decrease as the overtone frequencies increase.

Fundamentals of Communication Theory U. Pillai & A. Patel

Parseval's Theorem and Power Distribution

$$f(t) = \sum_{k=-\infty}^{\infty} c_k e^{jk\omega_0 t}$$

and

$$P = \frac{1}{T}\int_0^T |f(t)|^2\, dt = \sum_{k=-\infty}^{\infty} |c_k|^2 < \infty \tag{1.15}$$

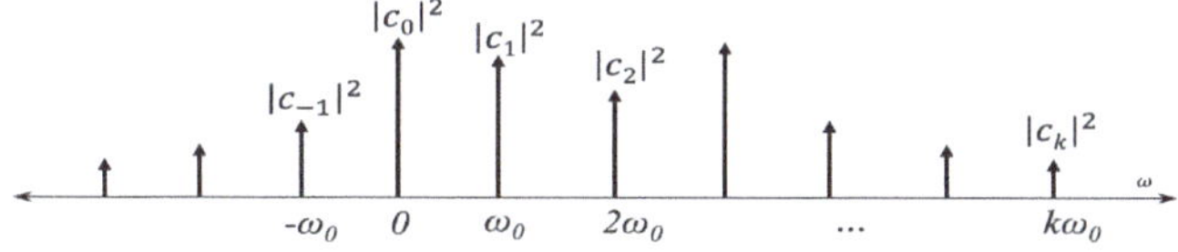

Figure 1.4: Power Spectrum

where $|c_k|^2$ is the signal power at frequency $k\omega_0$. From (1.15), for finite power ($P < \infty$) periodic signals $c_k \to 0$ as $k \to 0$.

Fourier Series Fourier Transforms Amplitude Modulation Angle Modulation Digital Modulation Techniques

Parseval's Theorem and Power Distribution

The left side in (1.15) represents the power in the time domain whereas the right side represents the power distribution over all discrete frequencies present in the signal.
Thus, the total power over all frequencies is the same as that over any period T in the time domain.

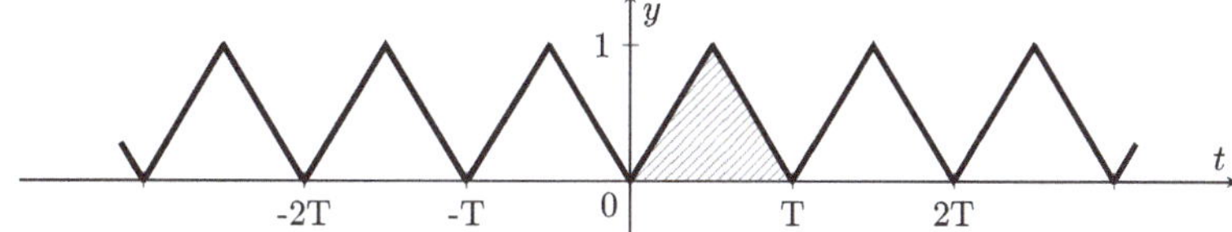

Figure 1.5: Periodic signal

Fundamentals of Communication Theory U. Pillai & A. Patel

Energy of the Error Signal

Define the error signal $e(t)$ as

$$e(t) = f(t) - \sum_{k=-\infty}^{\infty} c_k e^{jk\omega_0 t} \tag{1.16}$$

Using (1.15), we also get

$$\frac{1}{T}\int_0^T e(t)^2\,dt = 0 \tag{1.17}$$

This implies that the error signal has zero power or $e(t)$ is essentially flat everywhere, indicating the right side in (1.3) is a true representation of the left side atleast in the mean square sense.

Fourier Series Fourier Transforms Amplitude Modulation Angle Modulation Digital Modulation Techniques

Fourier Series Expansion of Real Periodic Signals

Real Periodic Signals ($f(t) = f^*(t)$)

$$f(t) = f(t+T) = f^*(t) = \sum_{k=-\infty}^{+\infty} c_k e^{jk\omega_0 t} \tag{1.18}$$

$$\begin{aligned} c_k &= \frac{1}{T}\int_0^T f(t)e^{-jk\omega_0 t}\,dt = \frac{1}{T}\int_0^T f(t)\left(\cos k\omega_0 t - j\sin k\omega_0 t\right)\,dt \\ &= \frac{1}{T}\int_0^T f(t)\cos k\omega_0 t\,dt - j\cdot\frac{1}{T}\int_0^T f(t)\sin k\omega_0 t\,dt = \alpha_k - j\beta_k \end{aligned} \tag{1.19}$$

where α_k is the real part and β_k is the imaginary part.

$$\alpha_k = \frac{1}{T}\int_0^T f(t)\cos(k\omega_0 t)\,dt \tag{1.20}$$

Fundamentals of Communication Theory U. Pillai & A. Patel

Fourier Series Expansion of Real Periodic Signals

Hence,

$$\alpha_{-k} = \frac{1}{T}\int_0^T f(t)\cos(-k\omega_0 t)\,dt = \alpha_k \tag{1.21}$$

Also,

$$\beta_k = \frac{1}{T}\int_0^T f(t)\sin(k\omega_0 t)\,dt \tag{1.22}$$

Hence,

$$\beta_{-k} = \frac{1}{T}\int_0^T f(t)\sin(-k\omega_0 t)\,dt = -\frac{1}{T}\int_0^T f(t)\sin(k\omega_0 t)\,dt = -\beta_k \tag{1.23}$$

$$\alpha_{-k} = \alpha_k, \quad \beta_{-k} = -\beta_k \tag{1.24}$$

Notice that, α_k, β_k retain even and odd properties. From (1.19)

$$c_k = \alpha_k - j\beta_k \tag{1.25}$$

Fourier Series Fourier Transforms Amplitude Modulation Angle Modulation Digital Modulation Techniques

Fourier Series Expansion of Real Periodic Signals

So that,

$$\begin{aligned} f(t) &= \sum_{k=-\infty}^{+\infty} c_k e^{jk\omega_0 t} = \sum_{k=-\infty}^{+\infty} (\alpha_k - j\beta_k)(\cos k\omega_0 t + j\sin k\omega_0 t) \quad \text{(real)} \\ &= \sum_{k=-\infty}^{+\infty} (\alpha_k \cos k\omega_0 t + \beta_k \sin k\omega_0 t) \\ &\qquad + j\left(\sum_{k=-\infty}^{+\infty} (\alpha_k \sin k\omega_0 t - \beta_k \cos k\omega_0 t)\right) \end{aligned} \tag{1.26}$$

If $f(t)$ is real, then the imaginary part in (1.26) must be identically equal to zero for all t.

Fundamentals of Communication Theory U. Pillai & A. Patel

Fourier Series Expansion of Real Periodic Signals

Thus, if $f(t)$ is real and periodic, from (1.26), then f(t) has the fourier representation

$$f(t) = \sum_{k=-\infty}^{+\infty} (\alpha_k \cos k\omega_0 t + \beta_k \sin k\omega_0 t) \tag{1.27}$$

$$\alpha_k = \frac{1}{T}\int_0^T f(t)\cos k\omega_0 t\, dt, \quad \beta_k = \frac{1}{T}\int_0^T f(t)\sin k\omega_0 t\, dt \tag{1.28}$$

or

$$f(t) = \sum_{k=-\infty}^{+\infty} A_k \cos(k\omega_0 t + \phi_k) \tag{1.29}$$

where, amplitude $A_k = \sqrt{\alpha_k^2 + \beta_k^2}$, and phase $\phi_k = \tan^{-1}\left(\frac{\beta_k}{\alpha_k}\right)$

Fourier Series Fourier Transforms Amplitude Modulation Angle Modulation Digital Modulation Techniques

Fourier Series Expansion of Real Periodic Signals

Since $\alpha_{-k} = \alpha_k, \beta_{-k} = -\beta_k \Rightarrow \beta_0 = 0$; if $f(t)$ is real, we can simplify (1.27) as

$$\begin{aligned} f(t) &= \sum_{k=-\infty}^{+\infty} (\alpha_k \cos k\omega_0 t + \beta_k \sin k\omega_0 t) \\ &= \alpha_0 + \sum_{k=1}^{\infty} \alpha_k \cos k\omega_0 t + \sum_{k=-1}^{-\infty} \alpha_k \cos k\omega_0 t \\ &+ \beta_0 + \sum_{k=1}^{\infty} \beta_k \sin k\omega_0 t + \sum_{k=-1}^{-\infty} \beta_k \sin k\omega_0 t \end{aligned} \tag{1.30}$$

Fundamentals of Communication Theory U. Pillai & A. Patel

Fourier Series Expansion of Real Periodic Signals

Substituting $k = -m$ for negative values of k,

$$\begin{aligned} f(t) = \alpha_0 + \sum_{k=1}^{\infty} \alpha_k \cos k\omega_0 t + \sum_{m=1}^{\infty} \alpha_{-m} \cos(-m\omega_0 t) \\ + \sum_{k=1}^{\infty} \beta_k \sin k\omega_0 t + \sum_{m=1}^{\infty} \beta_{-m} \sin(-m\omega_0 t) \end{aligned} \tag{1.31}$$

Using (1.24) and

$$\cos(-m\omega_0 t) = \cos m\omega_0 t, \quad \sin(-m\omega_0 t) = -\sin m\omega_0 t \tag{1.32}$$

Fourier Series Expansion of Real Periodic Signals

For real periodic signals $f(t)$, If $f(t)$ is a real signal,

$$f(t) = \alpha_0 + 2\left(\sum_{k=1}^{\infty} \alpha_k \cos k\omega_0 t + \sum_{k=1}^{\infty} \beta_k \sin k\omega_0 t\right) \tag{1.33}$$

where all α_k, β_k are real coefficients as in (1.28)

$$\begin{aligned} P &= \alpha_0^2 + 4\sum_{k=1}^{\infty} \alpha_k^2 \frac{1}{T}\int_0^T \cos^2 k\omega_0 t\, dt + 4\sum_{k=1}^{\infty} \beta_k^2 \int_0^T \sin^2 k\omega_0 t\, dt \\ &= \alpha_0^2 + 4\sum_{k=1}^{\infty} \alpha_k^2 \frac{1}{T}\int_0^T \frac{(1+\cos 2k\omega_0 t)}{2}\, dt \\ &\quad + 4\sum_{k=1}^{\infty} \beta_k^2 \frac{1}{T}\int_0^T \frac{(1-\cos 2k\omega_0 t)}{2}\, dt \\ &= \alpha_0^2 + 2\sum_{k=1}^{\infty} \alpha_k^2 + 2\sum_{k=1}^{\infty} \beta_k^2 \end{aligned} \tag{1.34}$$

Fourier Series Expansion of Real Signals

f(t) is real and even $\Rightarrow f(t) = f(-t)$

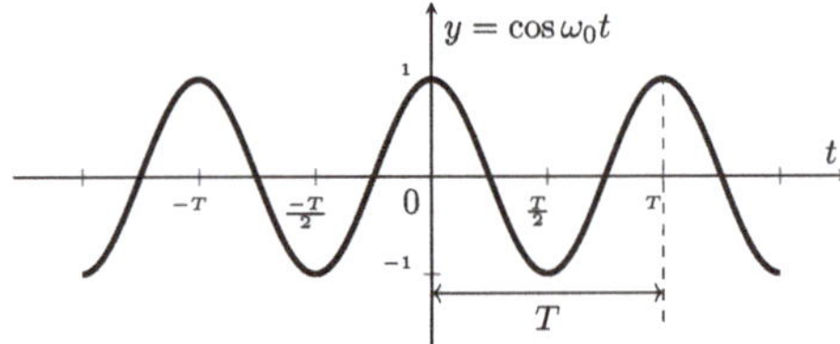

Figure 1.6: Real and even signal

$$f(-t) = f(t) = \sum_{k=-\infty}^{+\infty} (\alpha_k \cos k\omega_0 t + \beta_k \sin k\omega_0 t) \tag{1.35}$$

$$f(t) = \sum_{k=-\infty}^{+\infty} \{\alpha_k \cos k\omega_0(-t) + \beta_k \sin k\omega_0(-t)\} \tag{1.36}$$

Fourier Series Fourier Transforms Amplitude Modulation Angle Modulation Digital Modulation Techniques

Fourier Series Expansion of Real Periodic Signals

$$f(t) = \sum_{k=-\infty}^{+\infty} \{\alpha_k \cos k\omega_0(-t) - \beta_k \sin k\omega_0(t)\} \tag{1.37}$$

Adding (1.35) and (1.37), we obtain

$$f(t) = \sum_{k=-\infty}^{+\infty} \alpha_k \cos k\omega_0 t = \alpha_0 + 2\sum_{k=1}^{\infty} \alpha_k \cos k\omega_0 t \tag{1.38}$$

where

$$\alpha_k = \frac{1}{T}\int_{-T/2}^{T/2} f(t)\cos k\omega_0 t\, dt = \frac{2}{T}\int_{0}^{T/2} f(t)\cos k\omega_0 t\, dt \tag{1.39}$$

and from (1.34), the power relation for a real and even signal becomes

$$P = \frac{1}{T}\int_0^T f^2(t)\, dt = \alpha_0^2 + 2\sum_{k=1}^{\infty} \alpha_k^2 \tag{1.40}$$

Fundamentals of Communication Theory U. Pillai & A. Patel

Fourier Series Expansion of Real Periodic Signals

f(t) is real and odd $\Rightarrow f(-t) = -f(t)$ or $f(t) = -f(-t)$

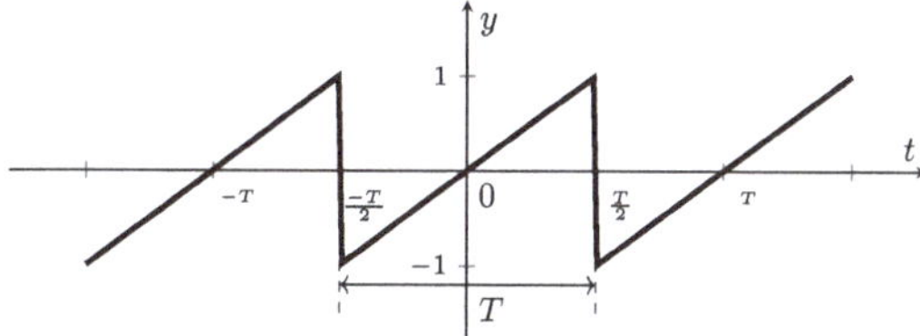

Figure 1.7: Real and odd signal

$$f(t) = \sum_{k=-\infty}^{+\infty} (\alpha_k \cos k\omega_0 t + \beta_k \sin k\omega_0 t) \tag{1.41}$$

$$-f(t) = f(-t) = -\left(\sum_{k=-\infty}^{+\infty} [\alpha_k \cos k\omega_0(-t) + \beta_k \sin k\omega_0(-t)] \right) \tag{1.42}$$

Fourier Series Expansion of Real Periodic Signals

$$-f(t) = -\sum_{k=-\infty}^{+\infty} (\alpha_k \cos k\omega_0 t - \beta_k \sin k\omega_0 t) \tag{1.43}$$

From (1.41) and (1.43)

$$2f(t) = 2\sum_{k=-\infty}^{+\infty} \beta_k \sin k\omega_0 t \tag{1.44}$$

$$f(t) = \sum_{k=-\infty}^{+\infty} \beta_k \sin k\omega_0 t = 2\sum_{k=1}^{\infty} \beta_k \sin k\omega_0 t. \tag{1.45}$$

where

$$\beta_k = \frac{1}{T}\int_{-T/2}^{T/2} f(t) \sin k\omega_0 t \, dt = \frac{2}{T}\int_{0}^{T/2} f(t) \sin k\omega_0 t \, dt \tag{1.46}$$

Since $\beta_0 = 0$, from (1.34), we get

$$P = \frac{1}{T}\int_0^T f^2(t)\, dt = 2\sum_{k=1}^{\infty} \beta_k^2. \tag{1.47}$$

Example 1.1

Square Wave signal (Even Periodic)

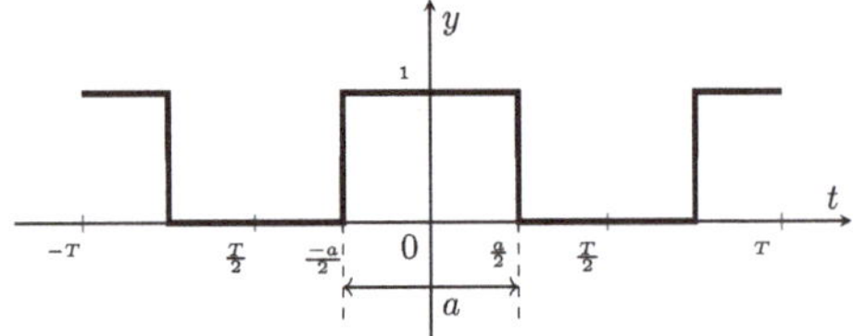

Figure 1.8: Even periodic square wave

$$
\begin{aligned}
c_k &= \frac{1}{T}\int_{-T/2}^{T/2} f(t)e^{-jk\omega_0 t}\,dt = \frac{1}{T}\int_{-a/2}^{a/2} 1\cdot e^{-jk\omega_0 t}\,dt \\
&= \frac{1}{T}\cdot\frac{e^{-jka\omega_0/2}-e^{jka\omega_0/2}}{-jk\omega_0} = \frac{2\sin(ka\omega_0/2)}{k\omega_0 T} = \frac{\sin(ka\omega_0/2)}{k\pi}
\end{aligned}
\tag{1.48}
$$

Fourier Series Fourier Transforms Amplitude Modulation Angle Modulation Digital Modulation Techniques

Example 1.1

Define

$$\operatorname{sinc} x = \frac{\sin x}{x} \tag{1.49}$$

then

$$c_k = \alpha_k = \frac{a\omega_0}{2\pi}\operatorname{sinc}\left(\frac{ka\omega_0}{2}\right) = \begin{cases} \frac{a\omega_0}{2\pi} & , \quad k=0 \\ \frac{a\omega_0}{2\pi}\operatorname{sinc}\left(\frac{ka\omega_0}{2}\right) & , \quad k=1,2,3,\ldots \end{cases} \tag{1.50}$$

Hence, from (1.38)

$$
\begin{aligned}
f(t) &= \alpha_0 + 2\sum_{k=1}^{\infty}\alpha_k \cos k\omega_0 t \\
&= \frac{a\omega_0}{2\pi}\left(1 + 2\sum_{k=1}^{\infty}\operatorname{sinc}\left(\frac{ka\omega_0}{2}\right)\cos k\omega_0 t\right)
\end{aligned}
\tag{1.51}
$$

Fundamentals of Communication Theory U. Pillai & A. Patel

Example 1.2

Special case: Even Periodic square wave with half duty cycle (a=T/2 in Example 1.1)

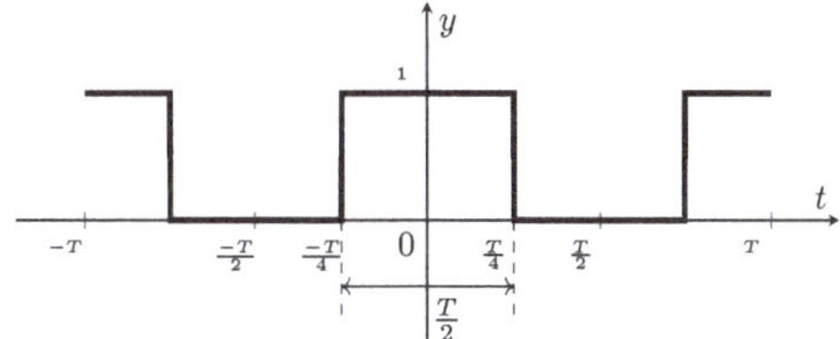

Figure 1.9: Even periodic square wave

$$f(t) = \alpha_0 + 2\sum_{k=1}^{\infty} \alpha_k \cos k\omega_0 t \tag{1.52}$$

$$\begin{aligned}\alpha_k &= \frac{a\omega_0}{2\pi}\,\text{sinc}\left(\frac{ka\omega_0}{2}\right) = \frac{\omega_0 T}{4\pi}\,\text{sinc}\left(\frac{k\omega_0 T}{4}\right)\\ &= \frac{2\pi}{4\pi}\,\text{sinc}\left(\frac{k2\pi}{4}\right) = \frac{1}{2}\,\text{sinc}\left(\frac{k\pi}{2}\right)\end{aligned} \tag{1.53}$$

Fourier Series Fourier Transforms Amplitude Modulation Angle Modulation Digital Modulation Techniques

Example 1.2

Hence,

$$\alpha_0 = \frac{1}{2} \tag{1.54}$$

$$\begin{aligned}\text{sinc}\left(\frac{k\pi}{2}\right) &= \frac{\sin(k\pi/2)}{k\pi/2}\\ &= \begin{cases} 0 & , \quad k = 2m, (m \text{ even}, \ m \neq 0)\\ \frac{\sin(2m+1)\pi/2}{(2m+1)\pi/2} = \frac{(-1)^m}{(2m+1)\pi/2} & , \quad k = 2m+1, (odd)\end{cases}\end{aligned} \tag{1.55}$$

Finally, Using (1.54)-(1.55) in (1.40), we get

$$f(t) = \frac{1}{2} + \frac{2}{\pi}\sum_{m=1}^{\infty} \frac{(-1)^m}{(2m+1)} \cos(2m+1)\omega_0 t \tag{1.56}$$

Fundamentals of Communication Theory U. Pillai & A. Patel

Example 1.3

Sawtooth Wave $f(t) = at, \ |t| \leq T/2$ (odd periodic)

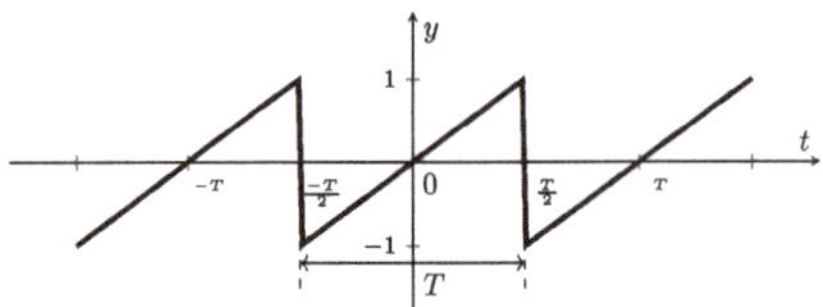

Figure 1.10: Periodic signal representation of y=at+b

$$f(t) = 2\sum_{k=1}^{\infty} b_k \sin k\omega_0 t \tag{1.57}$$

$$b_k = \frac{1}{T}\int_{-T/2}^{T/2} f(t)\sin k\omega_0 t\, dt = \frac{2}{T}\int_0^{T/2} f(t)\sin k\omega_0 t\, dt \tag{1.58}$$

For $t \in \left(-\frac{T}{2}, \frac{T}{2}\right)$,

$$f(t) = at \Rightarrow 1 = a.\frac{T}{2} \Rightarrow a = \frac{1}{T/2} \Rightarrow f(t) = \frac{t}{T/2} \tag{1.59}$$

Fourier Series Fourier Transforms Amplitude Modulation Angle Modulation Digital Modulation Techniques

Example 1.3

With (1.59) in (1.58), we get,

$$\begin{aligned} b_k &= \frac{2}{T}\int_0^{T/2} \frac{t}{(T/2)}\sin k\omega_0 t\, dt = \frac{4}{T^2}\int_0^{T/2} t\sin k\omega_0 t\, dt \\ &= \frac{4}{T^2}\left[t\cdot\frac{-\cos k\omega_0 t}{k\omega_0}\right]_0^{T/2} - \frac{4}{T^2}\int_0^{T/2}\frac{-\cos k\omega_0 t}{k\omega_0}\,dt \\ &= \frac{4}{T^2}\left[-\frac{T}{2}\cdot\frac{\cos(k\omega_0 T/2)}{k\omega_0} + \frac{\sin(k\omega_0 T/2)}{(k\omega_0)^2} - 0\right] \end{aligned} \tag{1.60}$$

Or with $T\omega_0 = 2\pi$,

$$b_k = -\frac{\cos(\pi k)}{\pi k} = -\frac{(-1)^k}{\pi k} = \frac{(-1)^{k+1}}{\pi k}, \quad k = 1, 2, 3, 4, \ldots \tag{1.61}$$

Fundamentals of Communication Theory U. Pillai & A. Patel

Example 1.3

$$f(t) = 2\sum_{k=1}^{\infty} b_k \sin k\omega_0 t = \frac{2}{\pi}\sum_{k=1}^{\infty}\frac{(-1)^{k+1}}{k}\sin k\omega_0 t$$
$$= \frac{2}{\pi}\left[\sin\omega_0 t - \frac{\sin 2\omega_0 t}{2} + \frac{\sin 3\omega_0 t}{3} - \cdots + (-1)^{k+1}\frac{\sin k\omega_0 t}{k} + \cdots\right] \tag{1.62}$$

Notice that power contained in the kth overtone is

$$b_k^2 = \frac{1}{T}\int_0^T \left(\frac{(-1)^{k+1}}{\pi k}\sin k\omega_0 t\right)^2 dt = \frac{1}{\pi^2 k^2 T}\int_0^T \left(\frac{1-\cos 2k\omega_0 t}{2}\right) dt$$
$$= \frac{1}{\pi^2 k^2} \tag{1.63}$$

Example 1.3

$$\text{Power, } P = \frac{1}{T}\int_{-T/2}^{T/2} |f(t)|^2\, dt = \frac{2}{T}\int_0^{T/2}\left(\frac{t}{T/2}\right)^2 dt$$
$$= \frac{8}{T^3}\left[\frac{t^3}{3}\right]_0^{T/2} = \frac{1}{3} \tag{1.64}$$

Using (1.48) and (1.61)

$$P = 2\sum_{k=1}^{\infty}|b_k|^2 = 2\sum_{k=1}^{\infty}\left[\frac{1}{\pi}\frac{(-1)^k}{k}\right]^2 = \frac{2}{\pi^2}\sum_{k=1}^{\infty}\frac{1}{k^2} \tag{1.65}$$

Equating (1.64) and (1.65), we get the well-known Euler's identity

$$\sum_{k=1}^{\infty}\frac{1}{k^2} = \frac{\pi^2}{6} \tag{1.66}$$

Example 1.4

Full Wave Sine Rectifier

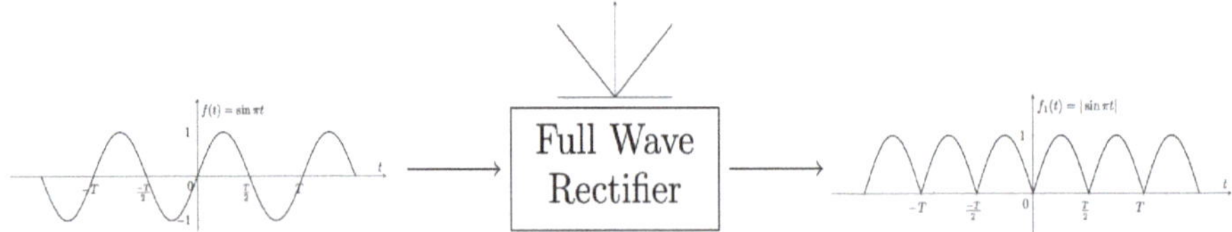

Figure 1.11: Full Wave Rectifier

Input signal,

$$f(t) = \sin \omega_0 t \tag{1.67}$$

is a pure harmonic, since it has a single frequency ω_0. Passing $f(t)$ through the full wave rectifier, we get (see Fig. 1.11)

$$f_1(t) = |\sin \omega_0 t| \tag{1.68}$$

Example 1.4

Notice that the input signal is a single tone odd periodic signal with period T while the output signal is an even periodic signal with period $T_1 = T/2$. Hence $\omega_1 = \frac{2\pi}{T_1} = \frac{4\pi}{T} = 2\omega_0$

$$\begin{aligned} a_0 &= \frac{1}{T_1}\int_0^{T_1} f_1(t)\,dt = \frac{2}{T}\int_0^{T/2} f(t)\,dt = \frac{2}{T}\int_0^{T/2} \sin \omega_0 t\,dt \\ &= -\frac{2}{T}\left[\frac{\cos \omega_0 t}{\omega_0}\right]_0^{T/2} = \frac{2}{T}\left[\frac{1-\cos(\omega_0 T/2)}{\omega_0}\right] = \frac{1}{\pi}(1-\cos \pi) = \frac{2}{\pi} \end{aligned} \tag{1.69}$$

$$a_k = \frac{1}{T_1}\int_0^{T_1} f(t) \cos k\omega_1 t\,dt = \frac{2}{T}\int_0^{T/2} \sin \omega_0 t \cdot \cos(2k\omega_0 t)\,dt \tag{1.70}$$

Since,

$$2\cos A \sin B = \sin(A+B) - \sin(A-B), \tag{1.71}$$

Example 1.4

Using (1.71) in (1.70), we get

$$\sin \omega_0 t \cdot \cos 2k\omega_0 t = \frac{1}{2} \sin(2k+1)\omega_0 t - \sin(2k-1)\omega_0 t \tag{1.72}$$

$$\begin{aligned}
a_k &= \frac{1}{T}\int_0^{T/2} [\sin\{(2k+1)\omega_0 t\} - \sin\{(2k-1)\omega_0 t\}]\, dt \\
&= \frac{1}{T}\left(-\frac{\cos(2k+1)\omega_0 t}{(2k+1)\omega_0}\bigg|_0^{T/2} + \frac{\cos(2k-1)\omega_0 t}{(2k-1)\omega_0}\bigg|_0^{T/2} \right) \\
&= \frac{1}{2\pi}\left(\frac{1-\cos\frac{(2k+1)\omega_0 T}{2}}{k+1} + \frac{\cos\frac{(2k-1)\omega_0 T}{2} - 1}{k-1} \right) \\
&= \frac{1}{2\pi}\left(\frac{1-\cos(2k+1)\pi}{2k+1} + \frac{\cos(2k-1)\pi - 1}{2k-1} \right) \\
&= \frac{1}{2\pi}\left(\frac{1-(-1)^{2k+1}}{2k+1} + \frac{(-1)^{2k+1} - 1}{2k-1} \right) = \frac{-2}{\pi(4k^2-1)}, \quad k = 1, 2, \ldots
\end{aligned} \tag{1.73}$$

Example 1.4

$$\begin{aligned}
f_1(t) = |\sin \omega_0 t| &= \frac{2}{\pi} - \frac{4}{\pi}\sum_{k=1}^{\infty} \frac{1}{(4k^2-1)} \cos 2k\omega_0 t \\
&= \frac{2}{\pi} - \frac{4}{\pi}\left(\frac{\cos 2\omega_0 t}{3} + \frac{\cos 4\omega_0 t}{15} + \cdots \right) \\
&= \frac{2}{\pi} - \frac{4}{\pi}\left[\sum_{k=1}^{\infty} \frac{\cos(2k\omega_0 t)}{(2k)^2 - 1} \right]
\end{aligned} \tag{1.74}$$

Since the amplitudes decay fast, we may approximate a full-wave rectifier using a finite number of terms as ($N \simeq 10$ or 20)

$$\hat{f}_1(t) = |\sin \omega_0 t| \simeq \frac{2}{\pi} - \frac{4}{\pi}\sum_{k=1}^{N} \frac{\cos(2k\omega_0 t)}{(2k)^2 - 1} \tag{1.75}$$

Notice that the output has several frequency components present as in (1.75).

Example 1.5

Half Wave Cosine Rectifier

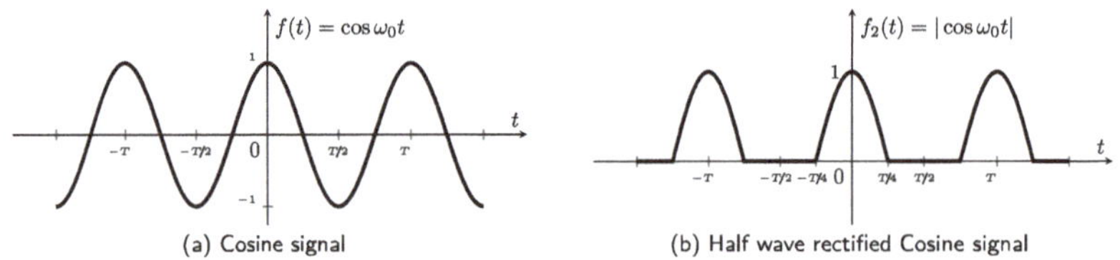

(a) Cosine signal (b) Half wave rectified Cosine signal

Figure 1.12: Half Wave Rectifier

Both input and output signals are even functions of time with the same period T. Using (1.20) and (1.38), from Fig. 1.12,

$$\begin{aligned} a_k &= \frac{1}{T}\int_{-T/4}^{T/4} f(t)\cos k\omega_0 t dt = \frac{2}{T}\int_0^{T} \cos\omega_0 t \cos k\omega_0 t dt \\ &= \frac{1}{T}\int_0^{T/4} [\cos(k+1)\omega_0 t + \cos(k-1)\omega_0 t] dt \end{aligned} \tag{1.76}$$

Fourier Series Fourier Transforms Amplitude Modulation Angle Modulation Digital Modulation Techniques

Example 1.5

$$\begin{aligned} a_k &= \frac{1}{T}\left[\frac{\sin(k+1)\omega_0 t}{(k+1)\omega_0} + \frac{\sin(k-1)\omega_0 t}{(k-1)\omega_0}\right]_0^{T/4} \\ &= \frac{1}{2\pi}\left(\frac{\sin(k+1)\pi/2}{k+1} + \frac{\sin(k-1)\pi/2}{k-1}\right) \\ &= \begin{cases} 0 & , \quad k = odd \\ \frac{1}{2\pi}\left(\frac{\sin(2m+1)\pi/2}{2m+1} + \frac{\sin(2m-1)\pi/2}{2m-1}\right) & , \quad k = even (k = 2m) \end{cases} \end{aligned} \tag{1.77}$$

For k = 2m;

$$a_{2m} = \frac{1}{2\pi}\left(\frac{(-1)^m}{2m+1} + \frac{(-1)^{m-1}}{2m-1}\right) = \frac{(-1)^{m-1}}{\pi(4m^2-1)} \tag{1.78}$$

with $a_0 = \frac{1}{\pi}$. Thus for a half wave cosine rectifier, we have

$$f(t) = \frac{1}{\pi} + \frac{1}{\pi}\sum_{k=1}^{\infty} \frac{(-1)^{k-1}}{(4k^2-1)} \cos(2k\omega_0 t) \tag{1.79}$$

Example 1.6

Full Wave Cosine Rectifier

$f(t)$ is periodic with period $T_1 = T/2$ and $\omega_1 = \frac{2\pi}{T_1} = \frac{4\pi}{T} = 2\omega_0$

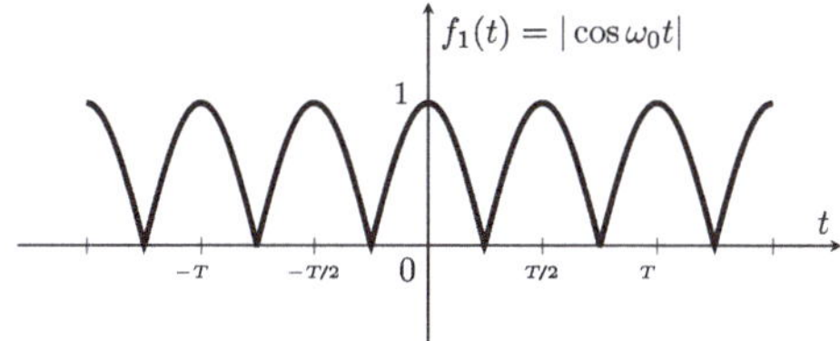

Figure 1.13: Full wave rectified Cosine signal

$$a_0 = \frac{2}{T_1}\int_0^{T_1/2} f(t)\,dt = \frac{4}{T}\int_0^{T/4} \cos\omega_0 t\,dt = \frac{4}{T}\cdot\frac{\sin\omega_0 T/4}{\omega_0} = \frac{2}{\pi} \tag{1.80}$$

Fourier Series Fourier Transforms Amplitude Modulation Angle Modulation Digital Modulation Techniques

Example 1.6

Using (1.39)

$$\begin{aligned}
a_k &= \frac{2}{T_1}\int_0^{T_1/2} f(t)\cos k\omega_1 t\,dt = \frac{4}{T}\int_0^{T/4} \cos\omega_0 t\,\cos 2k\omega_0 t\,dt \\
&= \frac{2}{T}\int_0^{T/4} [\cos(2k+1)\omega_0 t + \cos(2k-1)\omega_0 t]\,dt \\
&= \frac{2}{T}\left(\frac{\sin(2k+1)\omega_0 t}{(2k+1)\omega_0} + \frac{\sin(2k-1)\omega_0 t}{(2k-1)\omega_0}\right)\Bigg|_0^{T/2} \\
&= \frac{1}{\pi}\left(\frac{\sin(2k+1)\pi/2}{(2k+1)} + \frac{\sin(2k-1)\pi/2}{(2k-1)}\right) = \frac{1}{\pi}\left(\frac{(-1)^k}{2k+1} + \frac{(-1)^{k-1}}{2k-1}\right) \\
&= \frac{(-1)^k(2k-1-2k-1)}{\pi(4k^2-1)} = \frac{(-1)^{k+1}}{\pi(4k^2-1)}
\end{aligned} \tag{1.81}$$

Thus,

$$f_1(t) = \frac{2}{\pi} + \frac{2}{\pi}\sum_{k=1}^{\infty}(-1)^{k+1}\frac{\cos 2k\omega_0 t}{(2k)^2-1} \tag{1.82}$$

Example 1.7

Parabolic Periodic Signal

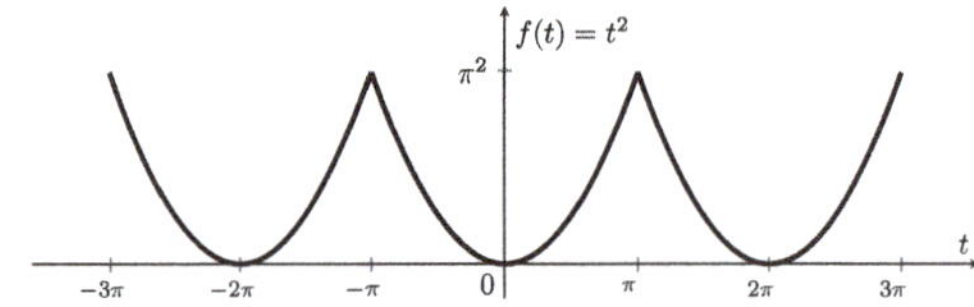

Figure 1.14: Parabolic periodic signal

$$a_0 = \frac{2}{\pi}\int_0^{\pi} x^2\,dx = \frac{2}{\pi}\frac{x^3}{3}\bigg|_0^{\pi} = \frac{2\pi^2}{3} \tag{1.83}$$

$$\begin{aligned} a_k &= \frac{2}{\pi}\int_0^{\pi} t^2\cos kt\,dt = \frac{2}{\pi}\left[\frac{t^2\sin kt}{k}\right]_0^{\pi} - \int_0^{\pi}\frac{2t\sin kt}{k}dt \\ &= \frac{-4}{\pi k}\left(\frac{-t\cos kt}{k}\bigg|_0^{\pi} + \underbrace{\int_0^{\pi}\frac{\cos kt}{k}dt}_{0}\right) = \frac{4\cos k\pi}{k^2} = \frac{(-1)^k}{4k^2} \end{aligned} \tag{1.84}$$

Fourier Series Fourier Transforms Amplitude Modulation Angle Modulation Digital Modulation Techniques

Example 1.7

For $t \in (-\pi, \pi)$,

$$f(t) = t^2 = \frac{2\pi^2}{3} - 4\left(\cos t - \frac{\cos 2t}{2^2} + \frac{\cos 3t}{3^2}\right) \quad , \quad |t| \le \pi \tag{1.85}$$

From (1.85), we get

$$f(0) = 0 = \frac{2\pi^2}{3} - 4\sum_{k=1}^{\infty}\frac{1}{k^2} \Longrightarrow \sum_{k=1}^{\infty}\frac{1}{k^2} = \frac{\pi^2}{6} \tag{1.86}$$

Fundamentals of Communication Theory U. Pillai & A. Patel

Example 1.8

Odd Square Wave

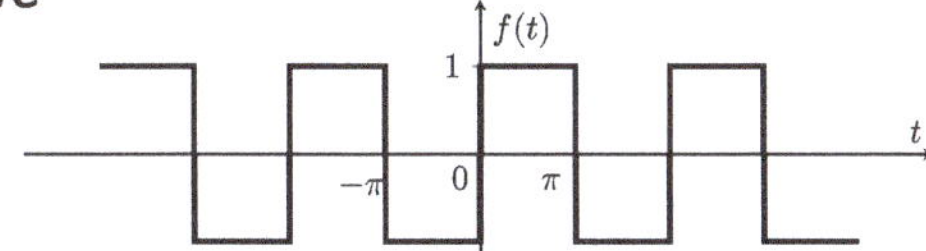

Figure 1.15: Periodic Square wave signal

From (1.41) and (1.45)-(1.46),

$$f(t) = 2\sum_{k=1}^{\infty} b_k \sin k\omega_0 t \tag{1.87}$$

$$\begin{aligned} b_k &= \frac{2}{\pi}\int_0^{\pi} 1\cdot \sin kt\, dt = \frac{2}{\pi}\frac{(-\cos kt)}{k}\bigg|_0^{\pi} = \frac{2}{\pi}(1-\cos k\pi) = \frac{2\left[1-(-1)^k\right]}{\pi} \\ &= \begin{cases} \frac{4}{\pi k} & , \quad k = odd \\ 0 & , \quad k = even \end{cases} \end{aligned} \tag{1.88}$$

Therefore,

$$f(t) = \frac{4}{\pi}\left(\sin t + \frac{\sin 3t}{3} + \frac{\sin 5t}{5} + \dots\right) \tag{1.89}$$

Gibb's Phenomenon

Fourier Series convergence at jump discontinuities

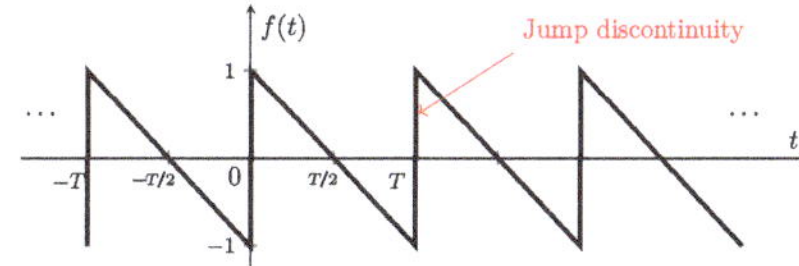

Figure 1.16: f(t) with jump discontinuity

Any jump discontinuity in a periodic wave can be removed using an appropriate sawtooth (ramp) waveform as shown below:

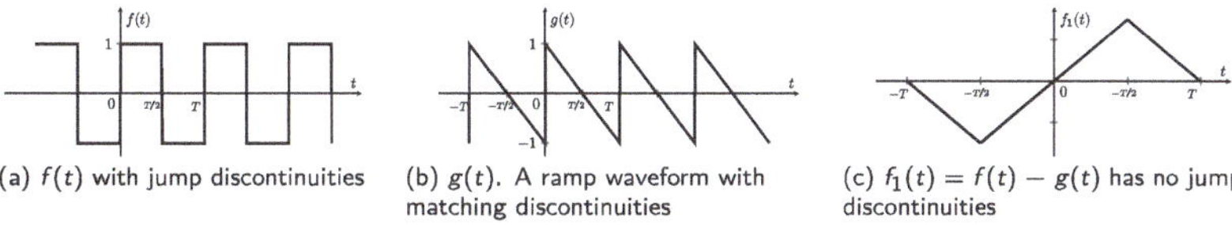

(a) $f(t)$ with jump discontinuities (b) $g(t)$. A ramp waveform with matching discontinuities (c) $f_1(t) = f(t) - g(t)$ has no jump discontinuities

Figure 1.17: Examples of Signals

Gibb's Phenomenon

The discontinuity in $f(t)$ has been removed by $g(t)$. Hence its convergence properties at jump discontinuities can be investigated by studying the convergence of the ramp function $g(t)$ at its discontinuities. $g(t)$ is an odd periodic function with $\omega_0 = \frac{2\pi}{T}$

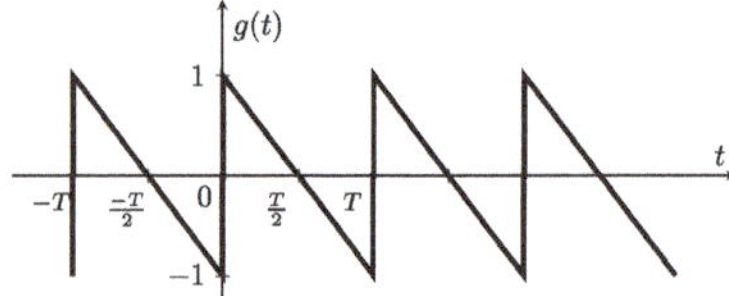

Figure 1.18: g(t) with jump discontinuity

$$g(t) = \left(1 - \frac{t}{T/2}\right) \quad ,0 < t < T/2 \tag{1.90}$$

so that using (1.45)

$$g(t) = 2\sum_{k=1}^{\infty} b_k \sin k\omega_0 t \tag{1.91}$$

Fourier Series Fourier Transforms Amplitude Modulation Angle Modulation Digital Modulation Techniques

Gibb's Phenomenon

where

$$\begin{aligned} b_k &= \frac{2}{T}\int_0^T g(t)\sin k\omega_0 t\,dt = \frac{2}{T}\int_0^{T/2}\left(1 - \frac{2t}{T}\right)\sin k\omega_0 t\,dt \\ &= \frac{2}{T}\left[-\frac{\cos k\omega_0 t}{k\omega_0}\bigg|_0^{T/2} - \frac{2}{T}\int_0^{T/2} t\sin k\omega_0 t dt\right] \\ &= \frac{1-\cos k\pi}{\pi k} - \frac{4}{T^2}\left[\frac{t(-\cos k\omega_0 t)}{k\omega_0}\bigg|_0^{T/2} + \int_0^{T/2}\frac{\cos k\omega_0 t}{k\omega_0}dt\right] \\ &= \frac{1-\cos\pi k}{\pi k} + \frac{\cos\pi k}{\pi k} + \underbrace{\frac{\sin k\omega_0 t)}{(k\omega_0)^2}\bigg|_0^{T/2}}_{0} = \frac{1}{\pi k} \quad ,k = 1, 2, \ldots \end{aligned} \tag{1.92}$$

Hence

$$g(t) = \left(1 - \frac{2t}{T}\right) = \frac{2}{\pi}\sum_{k=1}^{\infty}\frac{\sin k\omega_0 t}{k} \tag{1.93}$$

Fundamentals of Communication Theory U. Pillai & A. Patel

Gibb's Phenomenon

To investigate the convergence of (1.93) around the jump discontinuities $t = \pm kT$, consider the partial sum over n terms in (1.93). Thus

$$\begin{aligned} g_n(t) &= \frac{2}{\pi}\sum_{k=1}^{n}\frac{\sin k\omega_0 t}{k} = \frac{2\omega_0}{\pi}\sum_{k=1}^{n}\int_0^t \cos k\omega_0 x\, dx \\ &= \frac{4}{T}\int_0^t \sum_{k=1}^{n}\cos k\omega_0 x\, dx \end{aligned} \tag{1.94}$$

But,

$$\begin{aligned} \sum_{k=1}^{n}\cos k\omega_0 x &= \mathrm{Re}\left(\sum_{k=1}^{n} e^{jk\omega_0 x}\right) = \mathrm{Re}\left(e^{j\omega_0 x}\cdot\frac{1-e^{jn\omega_0 x}}{1-e^{j\omega_0 x}}\right) \\ &= \mathrm{Re}\left(e^{j(n+1)\omega_0 x/2}\cdot\frac{\sin(n\omega_0 x/2)}{\sin(\omega_0 x/2)}\right) \end{aligned} \tag{1.95}$$

Fourier Series Fourier Transforms Amplitude Modulation Angle Modulation Digital Modulation Techniques

Gibb's Phenomenon

$$\begin{aligned} \sum_{k=1}^{n}\cos k\omega_0 x &= \frac{\cos\left(\frac{(n+1)\omega_0 x}{2}\right)\sin\left(\frac{n\omega_0 x}{2}\right)}{\sin\left(\frac{\omega_0 x}{2}\right)} \\ &\simeq \frac{\cos\left(\frac{n\omega_0 x}{2}\right)\sin\left(\frac{n\omega_0 x}{2}\right)}{\sin\left(\frac{\omega_0 x}{2}\right)} = \frac{\sin n\omega_0 x}{2\sin\omega_0 x/2} \end{aligned} \tag{1.96}$$

Using (1.96) in (1.94), we get

$$g_n(t) = \frac{2}{T}\int_0^t \frac{\sin n\omega_0 x}{\sin\omega_0 x/2}\, dx \tag{1.97}$$

Let $n\omega_0 x = y \Longrightarrow dx = \frac{dy}{n\omega_0 x}$ and $\omega_0 x = \frac{y}{n}$, so that as $n \to \infty$, (1.97) becomes

$$\begin{aligned} g_n(t) &= \frac{2}{T}\int_0^{n\omega_0 t}\frac{\sin y}{\sin(y/2n)}\frac{dy}{n\omega_0} \simeq \frac{2}{n\omega_0 T}\int_0^{n\omega_0 t}\frac{\sin y}{(y/2n)}\, dy \\ &= \frac{2}{\pi}\int_0^{n\omega_0 t}\operatorname{sinc} y\, dy, \end{aligned} \tag{1.98}$$

where we have used $\sin(y/2n) \simeq y/2n$ as $n \to \infty$.

Gibb's Phenomenon

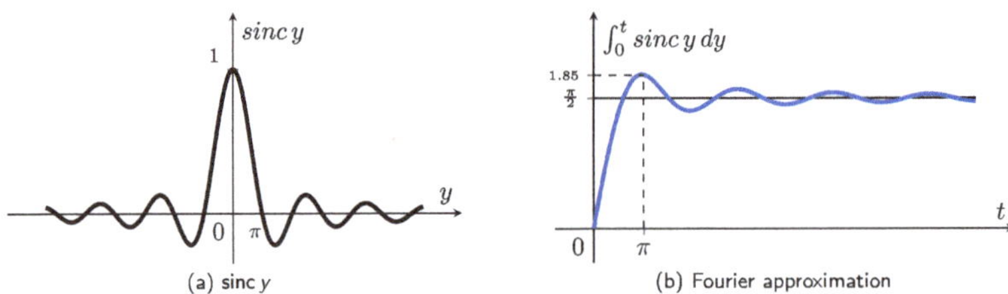

Figure 1.19: Fourier approximation near discontinuities

From the above figure, note that $\int_0^t \operatorname{sinc} y \, dy$ is maximum at $t = \pi$ and substituting this value into the upper limit in (1.98), we get the peak value occurs at

$$n\omega_0 t = \pi \Longrightarrow t_n = \frac{\pi}{n\omega_0} = \frac{T}{2n} \tag{1.99}$$

Hence $g_n(t)$ in (1.98) is maximized at the time instant $t = t_n = \frac{T}{2n}$. Notice that $t_n \to 0$ as the number of terms in the approximation in (1.94) increases, nevertheless the peak value persists.

Gibb's Phenomenon

Although $t_n = \frac{T}{2n} \to 0$, the corresponding maximum value at t_n tends to a finite value:

$$g_n(t_n) = \frac{2}{\pi} \int_0^{\pi} \operatorname{sinc} y \, dy \simeq \frac{1.85}{\pi/2} \simeq 1.179,$$

and it is independent of n, resulting in a fixed overshoot that is independent of n for every

$$\% \,\text{overshoot} = \frac{\max g_N(t) - g(0^+)}{g(0^+) - g(0^-)} = \frac{1.179 - 1}{1 - (-1)} = 8.95\% \simeq 9\% \tag{1.100}$$

i.e., For every periodic waveform with jump discontinuities, at every discontinuity, the Fourier Series summation results in an overshoot/ undershoot of approximately 9%. From Fig.(1.19), both the overshoots and undershoots tend to be steady, as $n \to \infty$ (as more and more terms are added). t_n moves closer to the discontinuity, but the overshoot remains steady independent of n.

Also see: Gibb's Phenomenon

Vibrato in Music

Vibrato is a musical phenomenon where experienced singers can modulate their voice that follows a sinusoidal pattern. Notice that $\cos\omega_0 t$ is a pure tone, whereas

$$\psi(t) = \cos(\omega_0 t + \beta \sin B_0 t) \tag{1.101}$$

represents a vibrato where the instantaneous frequency,

$$\omega(t) = \frac{d\psi(t)}{dt} = \frac{d}{dt}(\omega_0 t + \beta \sin B_0 t) = \omega_0 + \beta B_0 \cos \omega_0 t \tag{1.102}$$

varies with time in the frequency band $(\omega_0 - \beta B_0, \omega_0 + \beta B_0)$ sinusoidally.

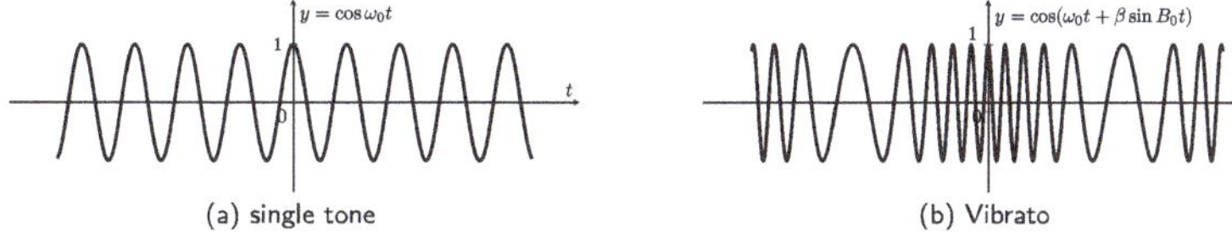

(a) single tone (b) Vibrato

Figure 1.20: Representation of vibrato signal

Fourier Series Fourier Transforms Amplitude Modulation Angle Modulation Digital Modulation Techniques

Vibrato in Music

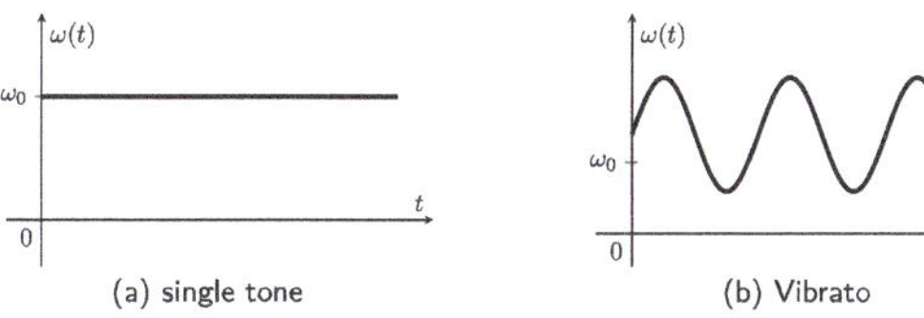

(a) single tone (b) Vibrato

Figure 1.21: Representation of vibrato signal

Frequency distribution of Vibrato

$$\begin{aligned} f(t) &= \cos(\omega_0 t + \beta \sin B_0 t) = \text{Re}\{e^{j(\omega_0 t + \beta \sin B_0 t)}\} \\ &= \text{Re}\{e^{j\omega_0 t} + e^{j\beta \sin B_0 t}\} = \text{Re}\{e^{j\omega_0 t} \cdot g(t)\} \end{aligned} \tag{1.103}$$

where

$$g(t) = e^{j\beta \sin B_0 t} \tag{1.104}$$

has period $T_0 = \frac{2\pi}{B_0}$. Hence $g(t)$ has a Fourier series expansion:

$$g(t) = \sum_{k=-\infty}^{\infty} a_k \, e^{jkB_0 t} \tag{1.105}$$

Vibrato in Music

where

$$a_k = \frac{1}{T_0}\int_{-T_0/2}^{T_0/2} g(t)\, e^{-jkB_0 t}\, dt = \frac{B_0}{2\pi}\int_{-\pi/B_0}^{\pi/B_0} e^{j(\beta \sin B_0 t - kB_0 t)}\, dt \quad (1.106)$$

Let $B_0 t = x$, $dt = \frac{dx}{B_0}$, so that (1.106) becomes

$$\begin{aligned} a_k &= \frac{B_0}{2\pi}\int_{-\pi}^{\pi} e^{j(\beta \sin x - kx)}\, \frac{dx}{B_0} = \frac{1}{2\pi}\int_{-\pi}^{\pi} e^{j(\beta \sin x - kx)}\, dx \\ &= \frac{1}{2\pi}\int_{-\pi}^{\pi} \cos(\beta \sin x - kx)\, dx = \frac{1}{\pi}\int_{0}^{\pi} \cos(\beta \sin x - kx)\, dx \\ &= J_k(\beta) \end{aligned} \quad (1.107)$$

Here, $J_k(\beta)$ represents the kth order Bessel function evaluated at β.
Also see: Bessel Function

Fourier Series Fourier Transforms Amplitude Modulation Angle Modulation Digital Modulation Techniques

Vibrato in Music

Using (1.107) in (1.104)-(1.105), we get

$$g(t) = e^{j\beta \sin B_0 t} = \sum_{k=-\infty}^{\infty} J_k(\beta)\, e^{jkB_0 t} \quad (1.108)$$

and substituting 1.108 in 1.103, we get

$$f(t) = \mathrm{Re}\left\{ \sum_{k=-\infty}^{\infty} J_k(\beta)\, e^{j(\omega_0 t + kB_0 t)} \right\} = \sum_{k=-\infty}^{\infty} J_k(\beta)\, \cos(\omega_0 + kB_0)t \quad (1.109)$$

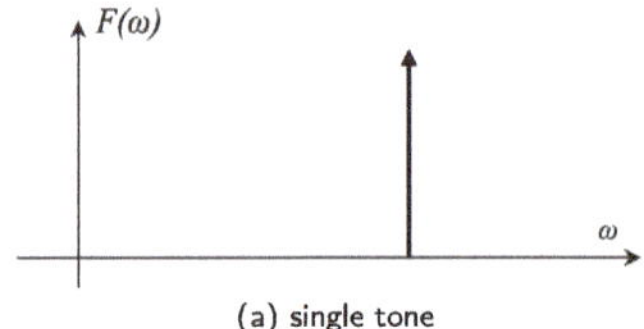

(a) single tone

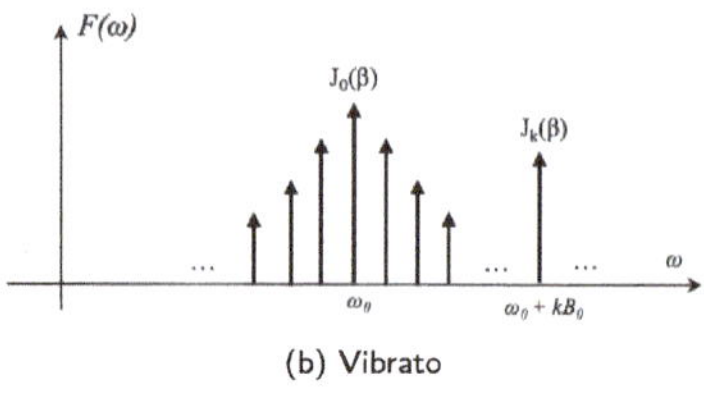

(b) Vibrato

Figure 1.22: Representation of vibrato signal

Fundamentals of Communication Theory U. Pillai & A. Patel

Vibrato in Music

Summary

$$\cos(\omega_0 t + \beta \sin B_0 t) = \sum_{k=-\infty}^{\infty} J_k(\beta) \cos(\omega_0 + kB_0)t$$
$$= J_0(\beta) \cos \omega_0 t + 2 \sum_{k=1}^{\infty} J_k(\beta) \cos(\omega_0 + kB_0)t \tag{1.110}$$

Vibrato generates simultaneous periodic frequency components centered at frequency $\omega_0, \omega_0 \pm B_0, \ldots, \omega_0 \pm kB_0$ that are in phase with varying amplitudes that are proportional to Bessel coefficients evaluated at some common value. The series on the right side in (1.110) can be approximated by about $\pm(\beta + 1)$ terms which preserver about 98% of its energy. Hence the effective bandwidth of a vibrato is about $-(\omega_0 + \beta B_0)$, $(\omega_0 + \beta B_0)$ or $2\beta B_0$ centered around ω_0. Interestingly, if we replace the $\sin B_0 t$ term in (1.101) with any other periodic waveform such as square wave (Example 1.8) or the ramp waveform (Fig.(1.18)), we can generate sounds such as fire alarm, ambulances etc.

Fourier Series Fourier Transforms Amplitude Modulation Angle Modulation Digital Modulation Techniques

Reversible Periodic Functions & Kepler's Equation

$$x = a \sin t \tag{1.111}$$

is a periodic function in t. Consider

$$y = t - x(t) = t - a \sin t \tag{1.112}$$

In (1.112), t is a continuously increasing function of y such that when $y = 0$, we get $t = 0$ and when $y = 2\pi$, t also increases to 2π. Hence $t - y = a \sin t$ can be expressed as an odd periodic function of either t or more importantly of y, and hence we can obtain the Fourier Series representation (see Fig.(1.23)) as

$$t - y = a \sin t = 2 \sum_{n=1}^{\infty} A_n \sin(ny) \tag{1.113}$$

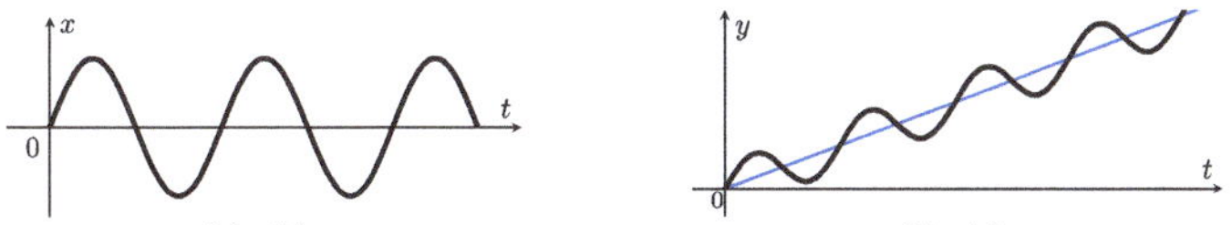

(a) $x(t)$

(b) $y(t)$

Figure 1.23: Reversible periodic functions

Fundamentals of Communication Theory U. Pillai & A. Patel

Reversible Periodic Functions & Kepler's Equation

where

$$
\begin{aligned}
A_n &= \frac{1}{2\pi}\int_0^{2\pi}(a\sin t)\sin ny\,dy \\
&= \underbrace{-\frac{1}{2\pi}\frac{\cos ny}{n}\cdot(a\sin t)\Big|_0^{2\pi}}_{0} + \frac{1}{2\pi n}\int_0^{2\pi}\cos ny\,d(a\sin t) \\
&= \frac{1}{2\pi n}\int_0^{2\pi}\cos ny\,d(t-y) = \frac{1}{2\pi n}\left[\int_0^{2\pi}\cos ny\,dt - \int_0^{2\pi}\cos ny\,dy\right] \\
&= \frac{1}{2\pi n}\int_0^{2\pi}\cos n(t - a\sin t)\,dt - \underbrace{\frac{1}{2\pi n}\frac{\sin ny}{n}\Big|_0^{2\pi}}_{0} \\
&= \frac{1}{2\pi n}\int_0^{2\pi}\cos(na\sin t - nt)\,dt = \frac{1}{n}J_n(na)
\end{aligned}
\tag{1.114}
$$

where $J_n(\beta)$ represents the nth order Bessel function in (1.107) evaluated at β.

Reversible Periodic Functions & Kepler's Equation

Hence it follows from 1.113 that

$$
t = y + \sum_{n=1}^{\infty}\frac{2}{n}J_n(na)\sin(ny) \tag{1.115}
$$

is the desired dual periodic representation of t in terms of y. The series in (1.115) is known as a Kapteyn series that converges rapidly when $a < 1$. (1.115) also represents solution to Kepler's problem of expressing the eccentric anomaly t in terms of the mean anomaly y in astronomy, that was originally solved by Bessel.
It is interesting to note that periodic functions as in (1.111), when linearly modified (see (1.112)) has a reversible periodic representation as in (1.115), where the time variable has a periodic representation along with a linear component in terms of the seasonal variable y, whether they be cherry blossoms or rain falls etc.

Reversible Periodic Functions & Kepler's Equation

More generally, if (1.111) is of the form

$$x(t) = \sum_{k=1}^{\infty} a_k \sin kt \quad , \tag{1.116}$$

then also (1.115) holds except that the coefficients in (1.113) will involve vibrato - type expansion.

2 Fourier Transforms

Abstract

Building upon the foundations of Fourier Series, this chapter introduces Fourier Transforms that extend the frequency representation from periodic signals to non-periodic signals. Properties essential for understanding signal behavior in the frequency domain, including Parseval's Theorem, Bessel's Equation, Bessel's Theorem, and Hilbert Transforms are introduced and analyzed here. Practical applications are reinforced through examples and detailed analysis of Fourier Transform properties. The discussions end up in Nyquist's Sampling Theorem that act as a bridge between the continuous and discrete signal domains, and forms the cornerstone of modern digital communication theory. Together, these concepts provide a comprehensive framework for analyzing and processing signals in the presence of real-world constraints.

U. Pillai and A. Patel, *Fundamentals of Communication Theory*,
https://doi.org/10.1007/978-3-032-20615-2_2

Table of Contents

1. Fourier Series

2. Fourier Transforms
2.1 Fourier's Theorem
2.2 Non-Periodic Signals
2.3 Parseval's Theorem
2.4 Bessel's Equation
2.5 Examples of Fourier Transforms
2.6 Bessel's Theorem
2.7 Fourier Transform Analysis
2.8 Properties of Fourier Transforms
2.9 Hilbert Transform
2.10 Sampling Theorem
3. Amplitude Modulation
4. Angle Modulation
5. Digital Modulation Techniques

Fourier Series | Fourier Transforms | Amplitude Modulation | Angle Modulation | Digital Modulation Techniques

Fourier's Theorem

Any periodic waveform with period $T \Rightarrow \left(\omega_0 = \frac{2\pi}{T}\right)$ has the representation:

$$f(t) = \sum_{k=-\infty}^{\infty} c_k e^{jk\omega_0 t} \tag{2.1}$$

If $f(t)$ is real, then as in (1.33)

$$f(t) = a_0 + 2\sum_{k=1}^{+\infty}(a_k \cos k\omega_0 t + b_k \sin k\omega_0 t) \tag{2.2}$$

Thus any real periodic signal can be decomposed in terms of a DC term and sine and cosine terms involving a fundamental frequency $\omega_0 = \frac{2\pi}{T}$ and its overtones $(k\omega_0, k = 2, 3, \ldots)$.

Fundamentals of Communication Theory U. Pillai & A. Patel

Fourier's Theorem

Also,

$$P = \frac{1}{T}\int_0^T |f(t)|^2\,dt = \sum_{k=-\infty}^{\infty} |c_k|^2 = \sum_{k=-\infty}^{\infty} \left(|a_k|^2 + |b_k|^2\right) \tag{2.3}$$

Thus, a signal that is periodic in the time domain has discrete representation in the frequency domain; i.e., energy of a periodic signal is distributed over a fundamental frequency ω_0 and its multiples $k\omega_0, k = 2, 3, \ldots$ and at no other frequencies.

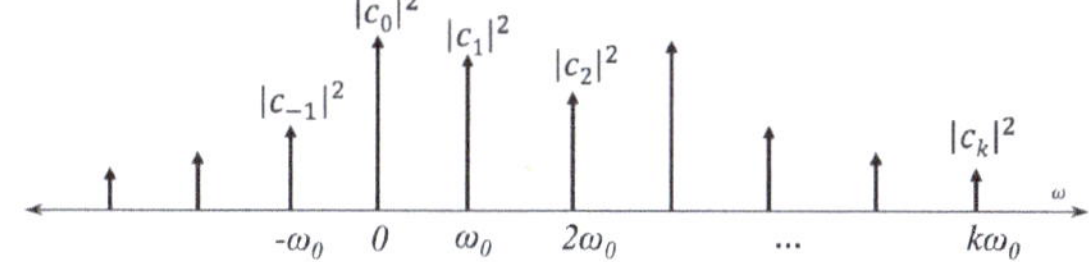

Figure 2.1: Power Spectrum

Non-Periodic Signals

What about non-periodic signals? How is their energy distributed in the frequency domain?

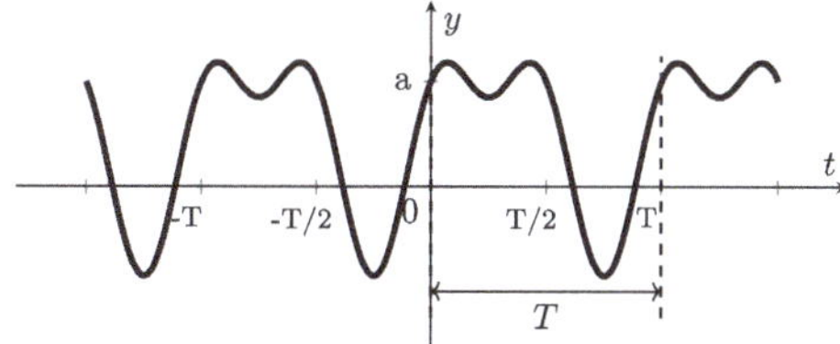

Figure 2.2: A Periodic Signal

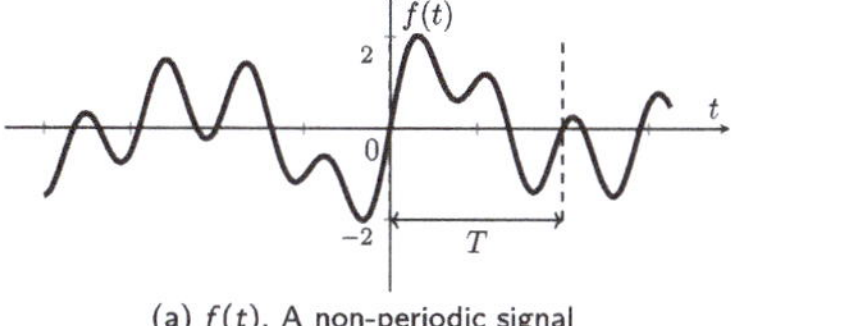

(a) $f(t)$, A non-periodic signal

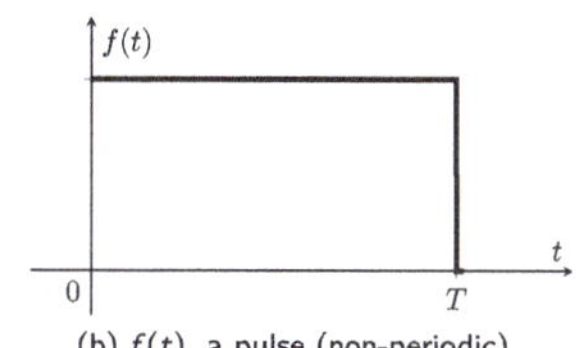

(b) $f(t)$, a pulse (non-periodic)

Figure 2.3: Non-Periodic Signals

Non-Periodic Signals

To understand this, it is best to generate a periodic signal $f_T(t)$ using a segment of duration T from a non-periodic signal $f(t)$ (from Fig. 2.3a) and analyze its properties. Then assume $T \to \infty$ to study the non-periodic signal.

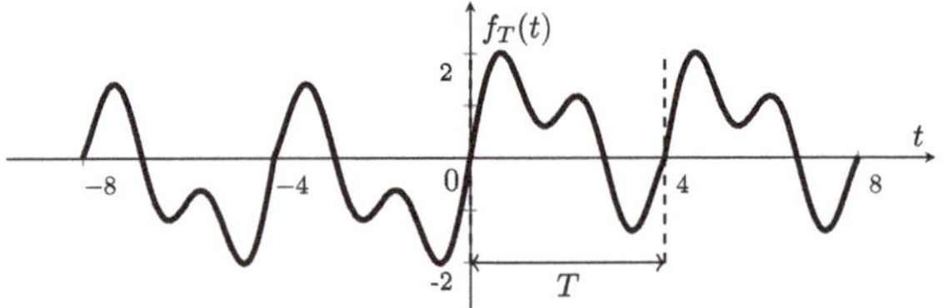

Figure 2.4: Generated Periodic Signal

$f_T(t)$ in Fig. 2.4 is periodic signal with period T. Hence has a Fourier series representation

Fourier Series Fourier Transforms Amplitude Modulation Angle Modulation Digital Modulation Techniques

Non-Periodic Signals

$$f_T(t) = \sum_{k=-\infty}^{\infty} c_k e^{jk\omega_0 t}, \quad \omega_0 = \frac{2\pi}{T} \tag{2.4}$$

$$c_k = \frac{1}{T} \int_{-T/2}^{T/2} f_T(t) e^{-jk\omega_0 t}\, dt \tag{2.5}$$

Also, $f_T(t) = f(t)$ in $\frac{-T}{2} < t < \frac{T}{2}$. Hence

$$c_k = \frac{1}{T} \int_{-T/2}^{T/2} f(t) e^{-jk\omega_0 t}\, dt = \frac{\omega_0}{2\pi} \int_{-T/2}^{T/2} f(t) e^{-jk\omega_0 t}\, dt \tag{2.6}$$

Note that as $T \to \infty$, we have

$$T \to \infty \implies \frac{\Delta\omega}{2\pi} = \frac{\omega_0}{2\pi} = \frac{1}{T} \implies \omega_0 \to 0. \tag{2.7}$$

Non-Periodic Signals

$$\text{Periodic Signal} \xrightarrow{T\to\infty} \text{Non-periodic Signal}$$

$$\text{(discrete frequency only)} \xrightarrow{T\to\infty} \text{(all frequencies are present)}$$

$$c_k = \frac{\omega_0}{2\pi}\int_{-T/2}^{T/2} f(t)e^{-jk\omega_0 t}\,dt \tag{2.8}$$

As $T \to \infty$, $f_T(t) \to f(t)$ and

$$\omega_0 = \frac{2\pi}{T} \implies \Delta\omega \to 0$$

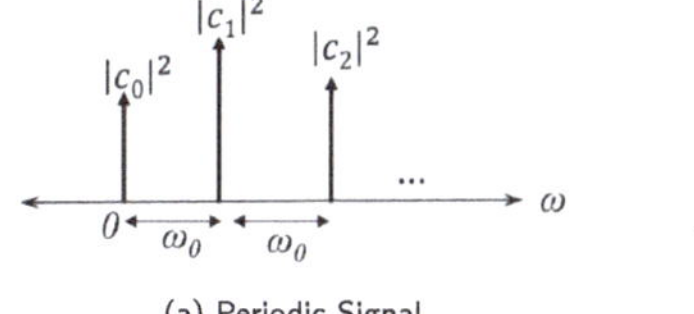

(a) Periodic Signal

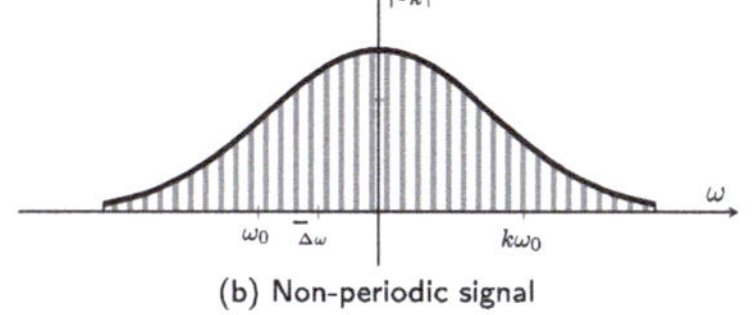

(b) Non-periodic signal

Figure 2.5: Frequency domain representation and Spectrum of a signal

Non-Periodic Signals

As $T \to \infty$, we get

$$\lim_{T\to\infty} c_k = \frac{\omega_0}{2\pi}\int_{-\infty}^{+\infty} f(t)e^{-jk\omega_0 t}\,dt \tag{2.9}$$

Motivated by (2.9), define

$$F(\omega) = \int_{-\infty}^{+\infty} f(t)e^{-j\omega t}\,dt, \tag{2.10}$$

where $F(\omega)$ is known as the Fourier Transform (F.T.) of $f(t)$.

Non-Periodic Signals

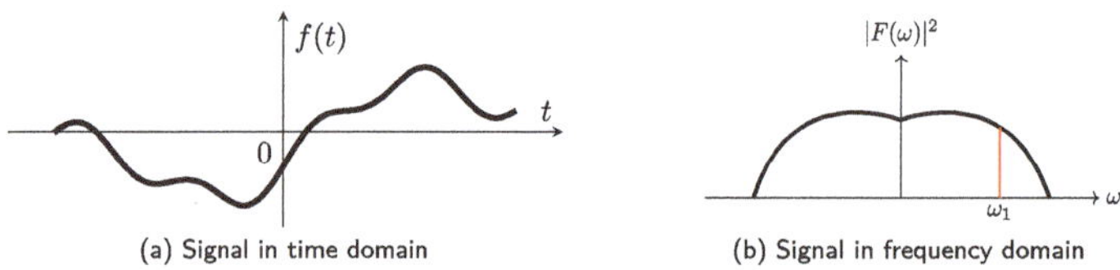

(a) Signal in time domain (b) Signal in frequency domain

Figure 2.6: Signal representation in time and frequency domain

Thus,

$$F(\omega) \overset{FT}{\Longleftrightarrow} \int_{-\infty}^{+\infty} f(t)e^{-j\omega t}\,dt \tag{2.11}$$

and

$$f(t) = \sum_{k=-\infty}^{+\infty} c_k e^{jk\omega_0 t} \tag{2.12}$$

and from (2.9) - (2.10)

$$c_k = \frac{\Delta\omega}{2\pi} F(k\Delta\omega) \tag{2.13}$$

Non-Periodic Signals

Substituting (2.13) into (2.12) and as $T \to \infty$, we get

$$f(t) = \frac{1}{2\pi} \sum_{k=-\infty}^{\infty} F(k\Delta\omega)e^{jk\Delta\omega t}\,\Delta\omega, \quad \omega = k\Delta\omega \tag{2.14}$$

or,

$$f(t) = \frac{1}{2\pi} \int_{-\infty}^{+\infty} F(\omega)e^{j\omega t}\,d\omega \tag{2.15}$$

Eq.(2.15) represents the inverse of the Fourier Transform(I.F.T.) and it recovers $f(t)$ from $F(\omega)$. Thus

$$F(\omega) = \int_{-\infty}^{+\infty} f(t)e^{-j\omega t}\,dt \tag{2.16}$$

and

$$f(t) = \frac{1}{2\pi} \int_{-\infty}^{+\infty} F(\omega)e^{j\omega t}\,d\omega \tag{2.17}$$

represent the Fourier Transform and its Inverse Transform pair.

Non-Periodic Signals

Time - Frequency

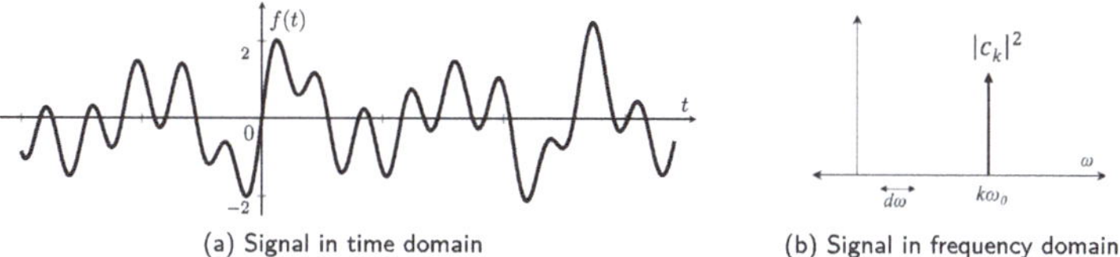

(a) Signal in time domain (b) Signal in frequency domain

Figure 2.7: Signal in time and frequency domain

Thus

$$f(t) = \lim_{T \to \infty} f_T(t) = \sum_{k=-\infty}^{\infty} \lim_{T \to \infty} c_k e^{jk\omega_0 t} \tag{2.18}$$

or

$$f(t) = \lim_{\Delta\omega \to \infty} \frac{1}{2\pi} \sum_{k=-\infty}^{\infty} F(k\Delta\omega) e^{jk\Delta\omega t} \Delta\omega \to \frac{1}{2\pi} \int_{-\infty}^{+\infty} F(\omega) e^{j\omega t}\, d\omega \tag{2.19}$$

Fourier Series | Fourier Transforms | Amplitude Modulation | Angle Modulation | Digital Modulation Techniques

Non-Periodic Signals

From (2.19), frequency to time (Inverse Fourier Transform) is given by

$$f(t) = \frac{1}{2\pi} \int_{-\infty}^{+\infty} F(\omega) e^{j\omega t}\, d\omega \tag{2.20}$$

Time to Frequency (Fourier Transform) is given by

$$F(\omega) = \int_{-\infty}^{+\infty} f(t) e^{-j\omega t}\, dt \tag{2.21}$$

In summary,

$$F(\omega) = \int_{-\infty}^{+\infty} f(t) e^{-j\omega t}\, dt \Longleftrightarrow f(t) = \frac{1}{2\pi} \int_{-\infty}^{+\infty} F(\omega) e^{j\omega t}\, d\omega \tag{2.22}$$

represent the fourier transform pair.

Fundamentals of Communication Theory U. Pillai & A. Patel

Non-Periodic Signals

For non-periodic signals, its total energy E is given by

$$E = \int_{-\infty}^{+\infty} |f(t)|^2 \, dt \tag{2.23}$$

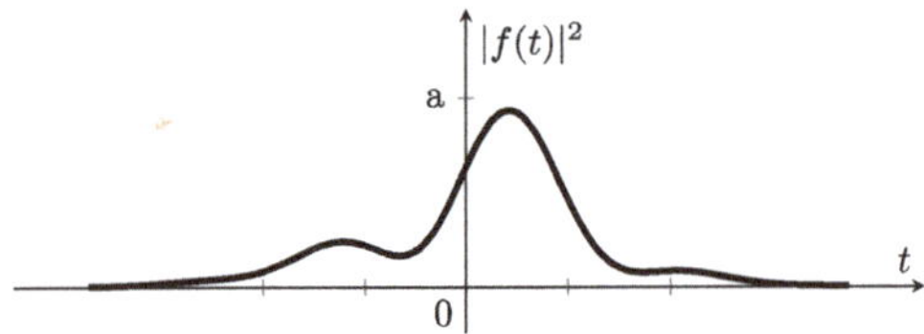

Figure 2.8: Time Energy Distributed Signal

How is this energy E distributed in the frequency domain?

Parseval's Theorem

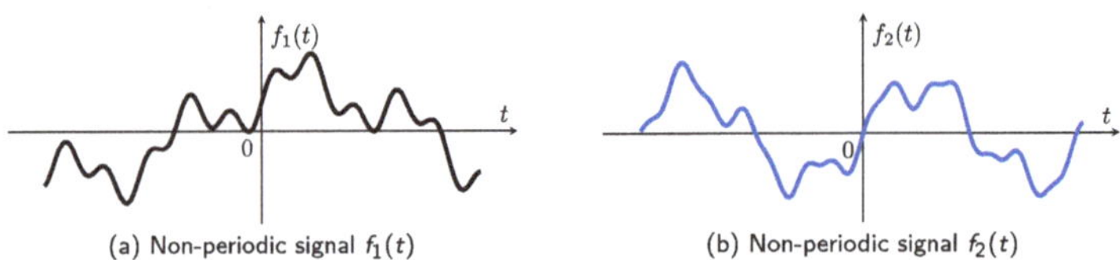

(a) Non-periodic signal $f_1(t)$ (b) Non-periodic signal $f_2(t)$

Figure 2.9: Non-Periodic Signals

Let $f_1(t)$ and $f_2(t)$ be two non-periodic signals. Then

$$F_1(\omega) = \int_{-\infty}^{+\infty} f_1(t)e^{-j\omega t} \, dt, \quad F_2(\omega) = \int_{\infty}^{+\infty} f_2(t)e^{-j\omega t} \, dt \tag{2.24}$$

$$f_1(t) = \frac{1}{2\pi} \int_{-\infty}^{+\infty} F_1(\omega)e^{j\omega t} \, d\omega, \quad f_2(t) = \frac{1}{2\pi} \int_{-\infty}^{+\infty} F_2(\omega)e^{j\omega t} \, d\omega \tag{2.25}$$

$$\int_{-\infty}^{+\infty} f_1(t)f_2^*(t) \, dt = \int_{-\infty}^{+\infty} f_1(t) \left(\frac{1}{2\pi} \int_{\infty}^{\infty} F_2^*(\omega)e^{-j\omega t} d\omega \right) dt \tag{2.26}$$

Parseval's Theorem

$$\begin{aligned}\int_{-\infty}^{+\infty} f_1(t)f_2^*(t)\,dt &= \int_{-\infty}^{+\infty} F_2^*(\omega)\left(\frac{1}{2\pi}\int_{-\infty}^{+\infty} f_1(t)e^{-j\omega t}\,dt\right)d\omega \\ &= \int_{-\infty}^{+\infty}\frac{1}{2\pi}F_2^*(\omega)F_1(\omega)d\omega\end{aligned} \tag{2.27}$$

or

$$\int_{-\infty}^{+\infty} f_1(t)f_2^*(t)\,dt = \frac{1}{2\pi}\int_{-\infty}^{+\infty} F_1(\omega)F_2^*(\omega)d\omega \tag{2.28}$$

i.e., the inner product of two time signals in the Time Domain is equal to their inner product in the Frequency Domain. Eq.(2.28) represents Parseval's Theorem.

Fourier Series | Fourier Transforms | Amplitude Modulation | Angle Modulation | Digital Modulation Techniques

Bessel's Equation

If $f_1(t) = f_2(t) = f(t)$, then we get

$$E = \int_{-\infty}^{+\infty} |f(t)|^2\,dt = \frac{1}{2\pi}\int_{-\infty}^{+\infty} |F(\omega)|^2\,d\omega \tag{2.29}$$

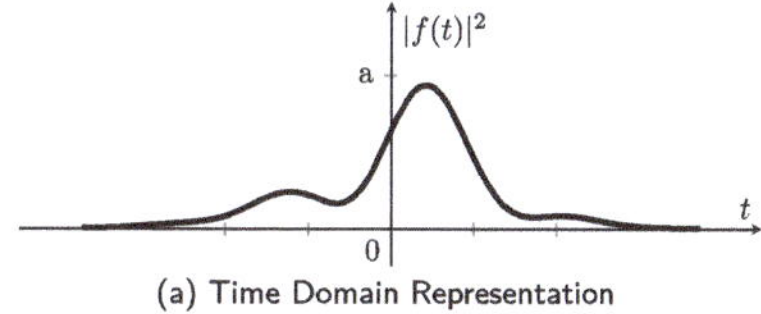

(a) Time Domain Representation

(b) Frequency Domain Representation

Figure 2.10: Energy signal in Time and Frequency domains

Fundamentals of Communication Theory U. Pillai & A. Patel

Bessel's Equation

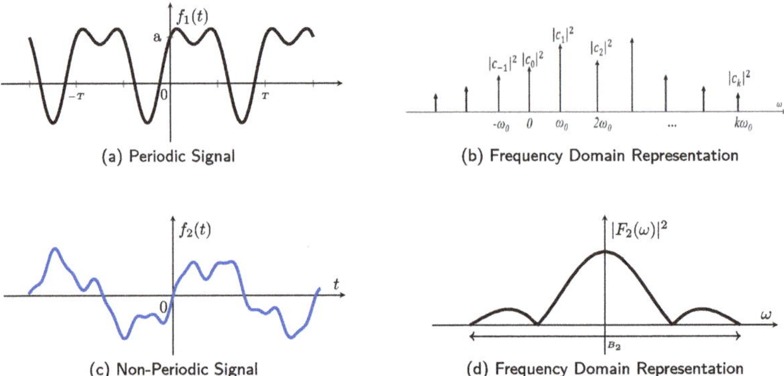

(a) Periodic Signal (b) Frequency Domain Representation

(c) Non-Periodic Signal (d) Frequency Domain Representation

Figure 2.11: Energy of a signal in frequency domain

Fourier Series Fourier Transforms Amplitude Modulation Angle Modulation Digital Modulation Techniques

Summary

Fourier Transform

$$\mathcal{F}\{f(t)\} = F(\omega) = \int_{-\infty}^{+\infty} f(t)e^{-j\omega t}\,dt \tag{2.30}$$

Inverse Fourier Transform

$$\mathcal{F}^{-1}\{F(\omega)\} = f(t) = \frac{1}{2\pi}\int_{-\infty}^{+\infty} F(\omega)e^{j\omega t}\,d\omega \tag{2.31}$$

$$E = \int_{-\infty}^{+\infty} |f(t)|^2\,dt > 0 = \int_{-\infty}^{+\infty} |F(\omega)|^2\,d\omega > 0 \tag{2.32}$$

Fundamentals of Communication Theory U. Pillai & A. Patel

Summary

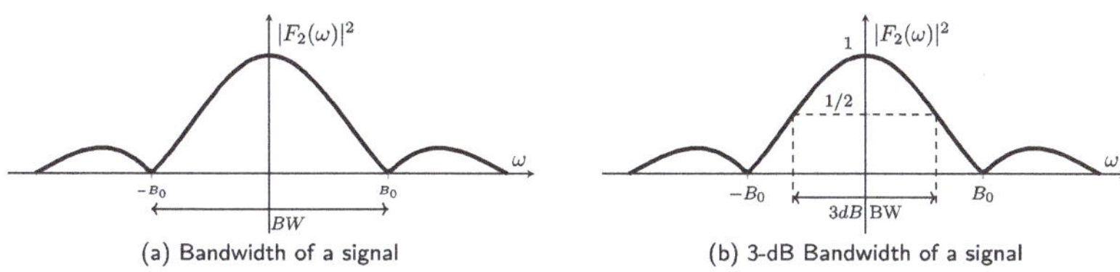

(a) Bandwidth of a signal (b) 3-dB Bandwidth of a signal

Figure 2.12: Signal Energy in frequency domain

Bandwidth 'BW' is where 'most' of the signal energy is concentrated in the frequency domain. A low pass signal has most of its energy around $\omega = 0$ region, and a high-pass signal has its energy in the higher frequency regions.
3-dB bandwidth is where the energy falls by half compared to the peak point (Fig. 2.12b).

Examples of Fourier Transforms

Rectangular Pulse

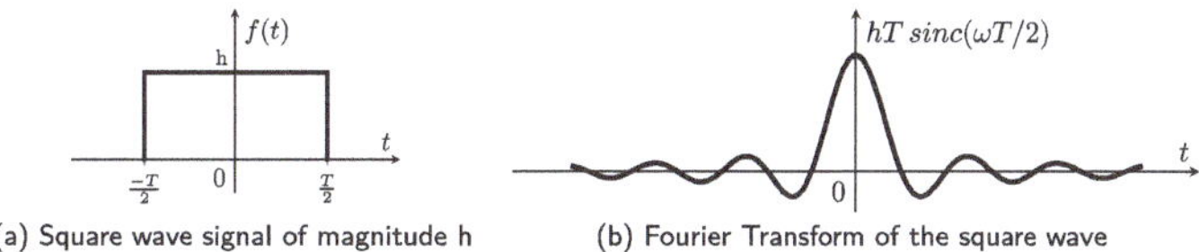

(a) Square wave signal of magnitude h (b) Fourier Transform of the square wave

Figure 2.13: Signal and its Fourier Transform

$$F(\omega) = \int_{-\infty}^{+\infty} f(t)e^{-j\omega t}\, dt = \int_{-T/2}^{T/2} he^{-j\omega t} dt = h\frac{e^{-j\omega t}}{j\omega}\bigg|_{-T/2}^{T/2}$$
$$= h\left(\frac{e^{-j\omega T/2} - e^{j\omega T/2}}{-j\omega}\right) = \frac{-2jh}{-j\omega}\sin(\omega T/2) = h\left(\frac{\sin(\omega T/2)}{\omega/2}\right) \tag{2.33}$$

$$f(t) \longleftrightarrow F(\omega) = hT\frac{\sin(\omega T/2)}{\omega T/2} = hT\,\text{sinc}\,(\omega T/2) \tag{2.34}$$

The Fourier Transform of an even rectangular wave is a sinc waveform.

Examples of Fourier Transform

When h=1, T=2;

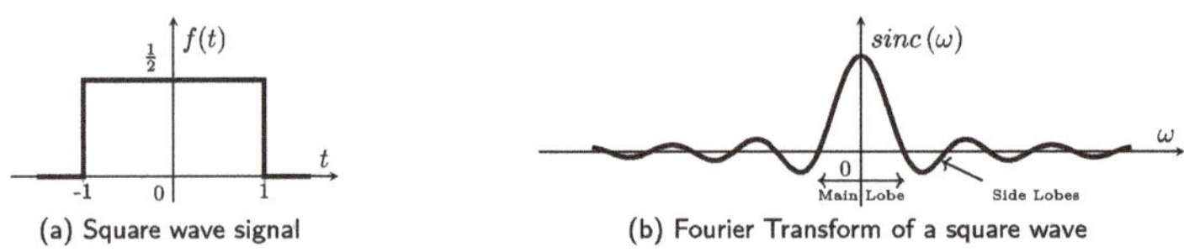

(a) Square wave signal

(b) Fourier Transform of a square wave

Figure 2.14: Signal and its Fourier Transform

$$E = \int_{-1}^{1} \underbrace{|f(t)|^2}_{\frac{1}{4}} dt = \frac{1}{2} = \frac{1}{2\pi} \int_{-\infty}^{+\infty} \text{sinc}^2 \, \omega \, d\omega \tag{2.35}$$

$$\int_{-\infty}^{+\infty} \text{sinc}^2 \, \omega \, d\omega = \pi \Longrightarrow \int_{-\infty}^{+\infty} \text{sinc}^2 \, t \, dt = \pi \Longrightarrow \int_{0}^{+\infty} \text{sinc}^2(t) \, dt = \frac{\pi}{2} \tag{2.36}$$

Examples of Fourier Transform

Bandwidth B: From (2.34), the first zero of $F(\omega)$ may be chosen as the signal bandwidth.

$$\text{sinc}\left(\frac{BT}{2}\right) = 0 = \frac{\sin(BT/2)}{BT/2} \tag{2.37}$$

$$\frac{BT}{2} = \pi \Longrightarrow B = \frac{2\pi}{T} \Longrightarrow f_0 = \frac{B}{2\pi} = \frac{1}{T} \tag{2.38}$$

One-sided $BW = B = \frac{1}{T}$

$$f_1(t) \leftrightarrow F_1(\omega) = \int_{-\infty}^{+\infty} f_1(t) e^{-j\omega t} dt \tag{2.39}$$

$$f_2(t) \leftrightarrow F_2(\omega) = \int_{-\infty}^{+\infty} f_2(t) e^{-j\omega t} dt \tag{2.40}$$

Bessel's Theorem

From (2.19),

$$y = \int_{-\infty}^{+\infty} f_1(t)f_2^*(t)dt = \frac{1}{2\pi}\int_{-\infty}^{+\infty} F_1(\omega)F_2^*(\omega)d\omega \tag{2.41}$$

$$E = \int_{-\infty}^{+\infty} |f(t)|^2 dt = \frac{1}{2\pi}\int_{-\infty}^{+\infty} |F(\omega)|^2 d\omega \tag{2.42}$$

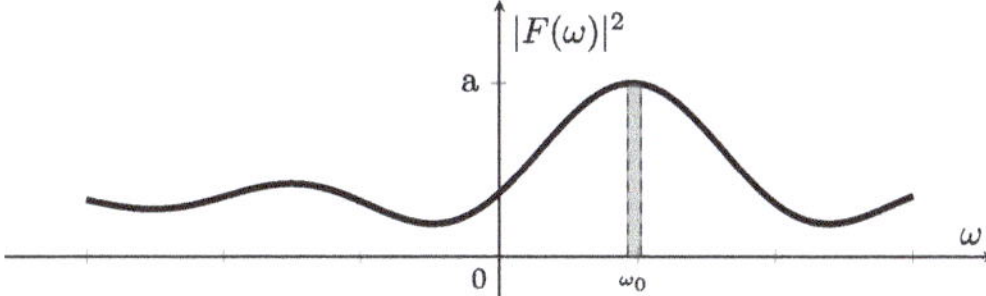

Figure 2.15: Energy of a signal in frequency domain

where the shaded region represents energy of $f(t)$ in $(\omega_0, \omega_0 + \Delta\omega)$

Fourier Series Fourier Transforms Amplitude Modulation Angle Modulation Digital Modulation Techniques

Fourier Transform Analysis

Fourier Transform of square wave with for the period [0,T]

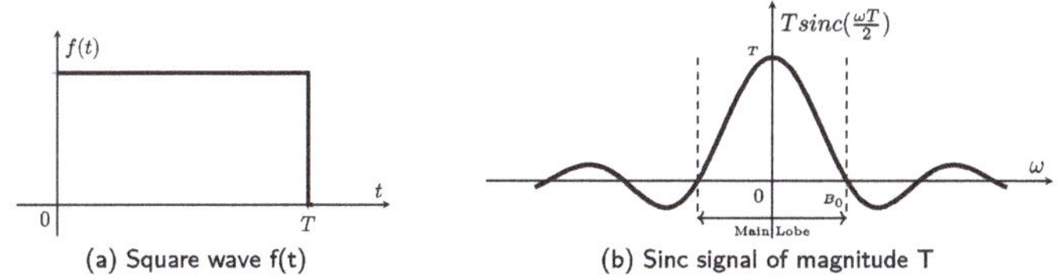

(a) Square wave f(t) (b) Sinc signal of magnitude T

Figure 2.16: Fourier Transform of a periodic square wave

$$\begin{aligned} F(\omega) &= Te^{-j\omega T/2} \cdot \frac{-2j\sin(\omega T/2)}{-j\omega T} = Te^{(-j\omega T/2)}\frac{\sin(\omega T/2)}{\omega T/2} \\ &= T \cdot sinc\,(\omega T/2) \cdot e^{-j\omega T/2} \end{aligned} \tag{2.43}$$

where, $T\,sinc\,(\omega T/2)$ represents the magnitude and $e^{-j\omega T/2}$ represents the phase of $F(\omega)$.

In Fig. 2.16b, about 90% of the signal energy is concentrated in main lobe.

Fundamentals of Communication Theory U. Pillai & A. Patel

Linearity

$$f_1(t) \longleftrightarrow F_1(\omega) \tag{2.44}$$

$$f_2(t) \longleftrightarrow F_2(\omega) \tag{2.45}$$

$$f(t) = a_1 f_1(t) + a_2 f_2(t) \tag{2.46}$$

$$\begin{aligned} F(\omega) &= \int_{-\infty}^{+\infty} f(t)e^{-j\omega t}dt = \int_{-\infty}^{+\infty} (a_1 f_1(t) + a_2 f_2(t))e^{-j\omega t}dt \\ &= a_1 \int_{-\infty}^{+\infty} f_1(t)e^{-j\omega t}dt + a_2 \int_{-\infty}^{+\infty} f_2(t)e^{-j\omega t}dt \\ &= a_1 F_1(\omega) + a_2 F_2(\omega) \end{aligned} \tag{2.47}$$

Fourier transform of a linear combination of signals is the same linear combination of their respective Fourier transforms.

Shift in Time

$$\sum_{k=1}^{n} a_k f_k(t) \longleftrightarrow \mathcal{F}\left(\sum_{k=-1}^{n} a_k f_k(t)\right) = \sum_{k=1}^{n} a_k F_k(\omega) \tag{2.48}$$

where

$$F_k(\omega) = \int_{-\infty}^{\infty} f_k(t)e^{-j\omega t}dt \tag{2.49}$$

Shift-in Time

$$\begin{aligned} f(t) &\longleftrightarrow F(\omega) \\ f_1(t) = f(t - t_0) &\longleftrightarrow \ ? \end{aligned} \tag{2.50}$$

$$F_1(\omega) = \int_{-\infty}^{\infty} f(t - t_0)e^{-j\omega t}dt \tag{2.51}$$

$$t - t_0 = \tau \Longrightarrow t = \tau + t_0 \Longrightarrow dt = d\tau \tag{2.52}$$

Therefore,

$$F_1(\omega) = \int_{-\infty}^{\infty} f(\tau)e^{-j\omega(\tau + t_0)}d\tau = e^{-j\omega t_0} \int_{-\infty}^{\infty} f(\tau)e^{-j\omega\tau}d\tau = e^{-j\omega t_0} F(\omega) \tag{2.53}$$

Shift in Time

$$\mathcal{F}\{f(t-t_0)\} = F(\omega)e^{-j\omega t_0} \tag{2.54}$$

$$f(t-t_0) \leftrightarrow F(\omega)e^{-j\omega t_0} \tag{2.55}$$

Shift-in-time results in a linear phase shift in frequency domain

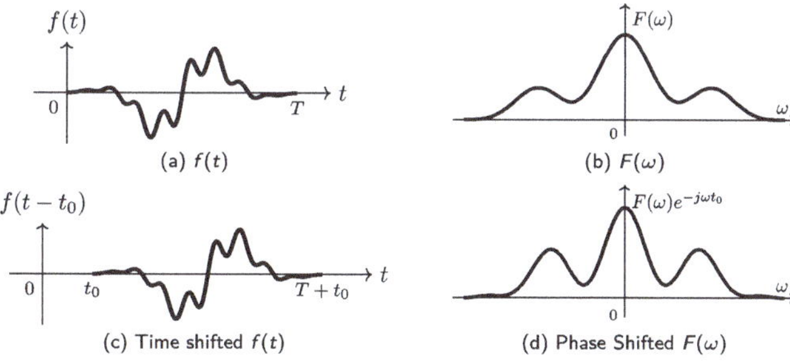

(a) $f(t)$ (b) $F(\omega)$

(c) Time shifted $f(t)$ (d) Phase Shifted $F(\omega)$

Figure 2.17: Shift in Time property of Fourier Transform

Fourier Series Fourier Transforms Amplitude Modulation Angle Modulation Digital Modulation Techniques

Transform of a Pulse Train

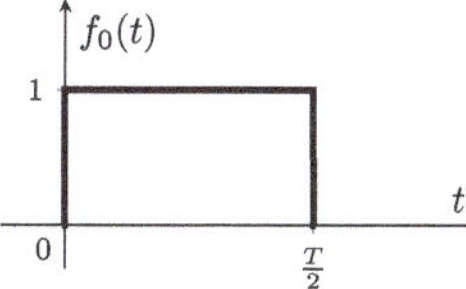

Figure 2.18: Rectangular pulse of magnitude 1

$$F_0(\omega) = \frac{T}{2}\sin(\omega T/4)e^{-j\omega T/4} \tag{2.56}$$

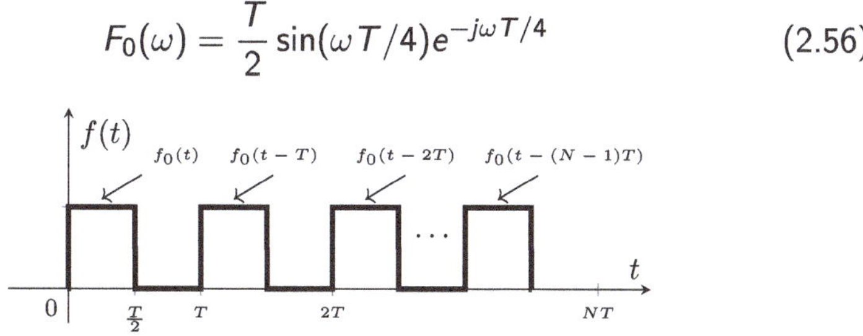

Figure 2.19: Rectangular pulse train of magnitude 1

Fundamentals of Communication Theory U. Pillai & A. Patel

Shift in Time

Radar Waveform (chain of pulses),

$$f(t) = \sum_{k=0}^{N-1} f_0(t - kT) \tag{2.57}$$

$$F(\omega) = \sum_{k=0}^{N-1} F_0(\omega) e^{-j\omega kT} = F_0(\omega) \sum_{k=0}^{N-1} e^{-j\omega kT} \tag{2.58}$$

$$S = 1 + r + r^2 + \cdots + r^{N-1}, \quad r = e^{-j\omega T} \tag{2.59}$$

$$rS = r + r^2 + \cdots + r^N \tag{2.60}$$

$$(1 - r)S = 1 - r^N \tag{2.61}$$

$$\begin{aligned} S &= \frac{1 - r^N}{1 - r} = \frac{1 - e^{-j\omega NT}}{1 - e^{-j\omega T}} = \frac{e^{-j\omega NT/2}}{e^{\ j\omega T/2}} \cdot \frac{(e^{j\omega NT/2} - e^{-j\omega NT/2})}{(e^{j\omega T/2} - e^{-j\omega T/2})} \\ &= e^{-j\omega(N-1)T/2} \cdot \frac{\sin(N\omega T/2)}{\sin(\omega T/2)} \end{aligned} \tag{2.62}$$

Shift in Time

$$\begin{aligned} F(\omega) &= F_0(\omega) \cdot \frac{\sin(N\omega T/2)}{\sin(\omega T/2)} \cdot e^{-j\omega T/2} \\ &= \frac{T}{2} \operatorname{sinc}(\omega T/4) \cdot \frac{\sin(N\omega T/2)}{\sin(\omega T/2)} \cdot e^{-j\omega(N+\frac{1}{2})T/2} \end{aligned} \tag{2.63}$$

where we have made use of (2.56)

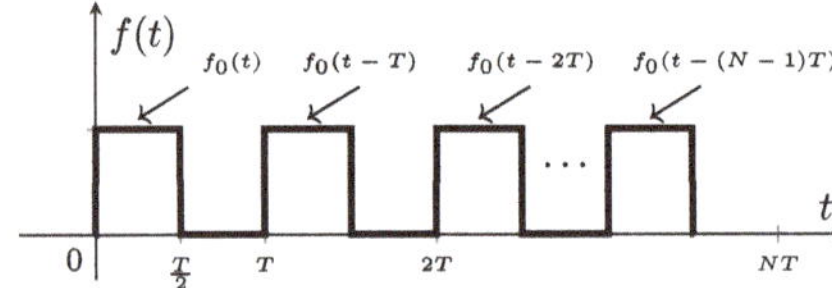

Figure 2.20: Rectangular pulse train of magnitude 1

$$\begin{aligned} F(\omega) &= \frac{T}{2} \cdot \frac{\sin(\omega T/4)}{(\omega T/4)} \cdot \frac{\sin(N\omega T/2)}{\sin(\omega T/2)} e^{-j(N+\frac{1}{2})\omega T/2} \\ &= NT \cdot \frac{\sin(N\omega T/2)}{\cos(\omega T/2)} e^{-j(N+\frac{1}{2})\omega T/2} \end{aligned} \tag{2.64}$$

Shift in Time

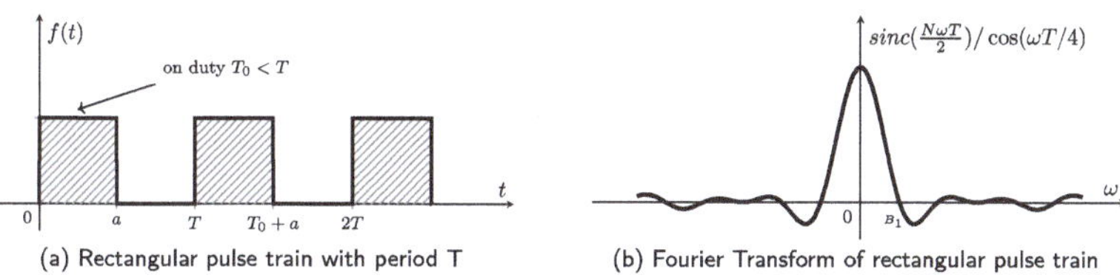

(a) Rectangular pulse train with period T

(b) Fourier Transform of rectangular pulse train

Figure 2.21: Signal and its Fourier Transform

Here,

$$\frac{a}{T} = \text{duty cycle} \tag{2.65}$$

$$B_1 = \frac{2\pi}{NT} = \frac{1}{2N} B_0 = \frac{B_0}{2N} \tag{2.66}$$

Fourier Series Fourier Transforms Amplitude Modulation Angle Modulation Digital Modulation Techniques

Shift in Time

Shift in Time Property

$$f(t - t_0) \longleftrightarrow F(\omega)e^{-j\omega t_0} \tag{2.67}$$

$$f(t + t_0) \longleftrightarrow F(\omega)e^{+j\omega t_0} \tag{2.68}$$

From (2.67) - (2.68) and using

$$\frac{f(t + t_0) + f(t - t_0)}{2} \longleftrightarrow F(\omega) \cos \omega t_0 \tag{2.69}$$

$$\frac{f(t + t_0) - f(t - t_0)}{2j} \longleftrightarrow F(\omega) \sin \omega t_0 \tag{2.70}$$

Fundamentals of Communication Theory U. Pillai & A. Patel

Shift in Frequency

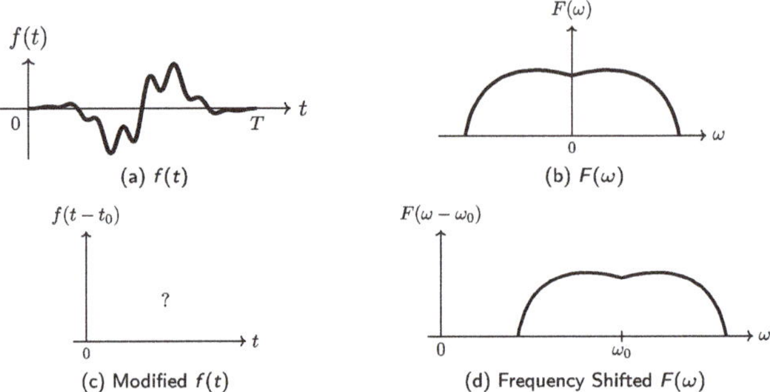

Figure 2.22: Shift in Frequency property of Fourier Transform

$$f(t) \longleftrightarrow F(\omega - \omega_0) \tag{2.71}$$

$$\mathcal{F}^{-1}\{F(\omega - \omega_0)\} = \frac{1}{2\pi}\int_{-\infty}^{+\infty} F(\omega - \omega_0)e^{j\omega t}d\omega \tag{2.72}$$

Fourier Series Fourier Transforms Amplitude Modulation Angle Modulation Digital Modulation Techniques

Shift in Frequency

Replacing $\omega - \omega_0$ with x in (2.72) gives:

$$f_1(t) = \frac{1}{2\pi}\int_{-\infty}^{+\infty} F(x)e^{j(x+\omega_0)t}\,dx = e^{j\omega_0 t}\underbrace{\frac{1}{2\pi}\int_{-\infty}^{+\infty} F(x)e^{jxt}\,dx}_{f(t)\text{ after replacing x with }\omega} \tag{2.73}$$

$$= e^{j\omega_0 t}\frac{1}{2\pi}\int_{-\infty}^{+\infty} F(\omega)e^{j\omega t}\,d\omega = f(t)e^{j\omega_0 t}$$

or

$$f(t)e^{j\omega_0 t} \longleftrightarrow F(\omega - \omega_0) \tag{2.74}$$

$$f(t)e^{-j\omega_0 t} \longleftrightarrow F(\omega + \omega_0) \tag{2.75}$$

Fundamentals of Communication Theory U. Pillai & A. Patel

Modulation Property

Adding and subtracting (2.98) to (2.75), we get (using linearity property)

$$f(t)\cos\omega_0 t \longleftrightarrow \frac{F(\omega-\omega_0)+F(\omega+\omega_0)}{2} \tag{2.76}$$

and

$$f(t)\sin\omega_0 t \longleftrightarrow \frac{F(\omega-\omega_0)+F(\omega-\omega_0)}{2j} \tag{2.77}$$

The time domain operation of multiplying by $\cos\omega_0 t$ in (2.76) is known as modulation and ω_0 is known as the carrier frequency.

Modulation in time domain results in frequency shift of the spectrum in the frequency domain by the carrier frequency.

Modulation Property

$$\mathcal{F}\{f(t)\cos\omega_0 t\} = \frac{1}{2}\left[F(\omega+\omega_0)+F(\omega-\omega_0)\right] \tag{2.78}$$

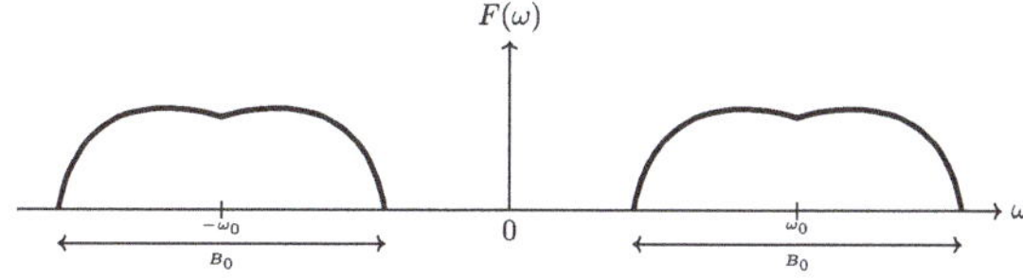

Figure 2.23: Cosine modulated f(t) in Frequency Domain

$$\mathcal{F}\{f(t)\sin\omega_0 t\} = \frac{j}{2}\left[F(\omega+\omega_0)-F(\omega-\omega_0)\right] \tag{2.79}$$

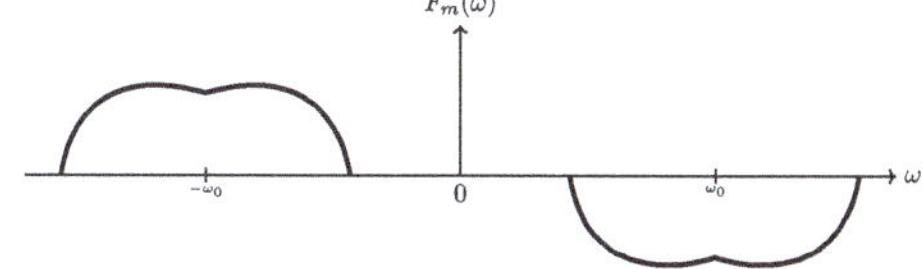

Figure 2.24: Sine modulated f(t) in Frequency Domain

Modulation Property

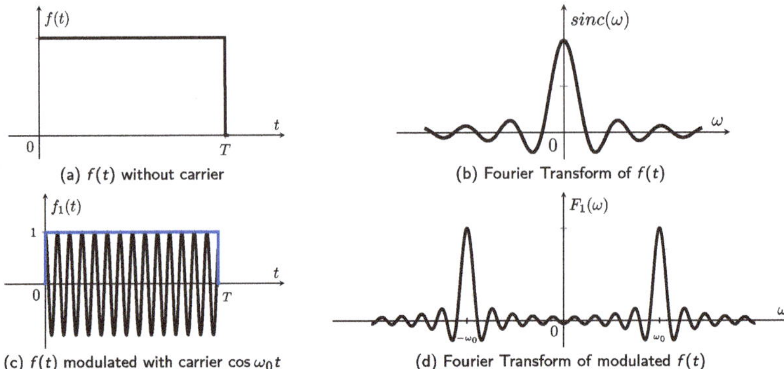

(a) $f(t)$ without carrier

(b) Fourier Transform of $f(t)$

(c) $f(t)$ modulated with carrier $\cos \omega_0 t$

(d) Fourier Transform of modulated $f(t)$

Figure 2.25: Modulation property of Fourier Transform

Fourier Series Fourier Transforms Amplitude Modulation Angle Modulation Digital Modulation Techniques

Modulation Property

$$f(t) \leftrightarrow F(\omega) \tag{2.80}$$

$$f(t)e^{j\omega_0 t} \leftrightarrow F(\omega - \omega_0) \tag{2.81}$$

$$f(t)e^{-j\omega_0 t} \leftrightarrow F(\omega + \omega_0) \tag{2.82}$$

Combining (2.81) and (2.82), we get,

$$f(t) \underbrace{\left[e^{j\omega_0 t} + e^{-j\omega_0 t}\right]}_{2\cos(\omega_0 t)} \longleftrightarrow F(\omega - \omega_0) + F(\omega + \omega_0) \tag{2.83}$$

$$g(t) = \underbrace{f(t)}_{\text{signal}} \cdot \underbrace{\cos(\omega_0 t)}_{\text{modulation}} \longleftrightarrow \frac{F(\omega - \omega_0) + F(\omega + \omega_0)}{2} \tag{2.84}$$

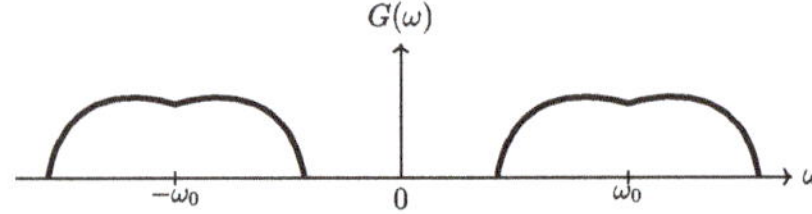

Figure 2.26: Fourier Transform of g(t)

Fundamentals of Communication Theory U. Pillai & A. Patel

Modulation Property

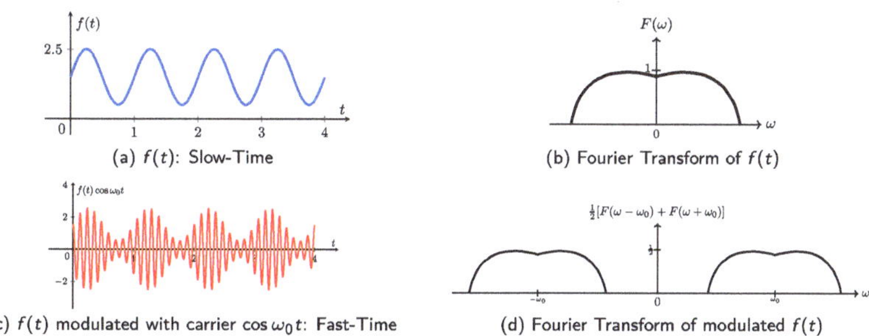

(a) $f(t)$: Slow-Time

(b) Fourier Transform of $f(t)$

(c) $f(t)$ modulated with carrier $\cos \omega_0 t$: Fast-Time

(d) Fourier Transform of modulated $f(t)$

Figure 2.27: Representation of Modulation property of Fourier Transform

Fourier Series Fourier Transforms Amplitude Modulation Angle Modulation Digital Modulation Techniques

Modulation Property

Find the Fourier Transform of a modulated pulse train shown below?

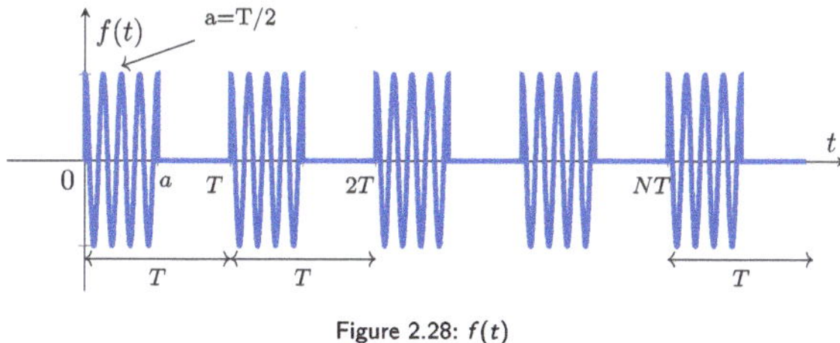

Figure 2.28: $f(t)$

Answer: Same as in (2.64), but shifted to left and right in the frequency domain by $\pm\omega_0$.

Fundamentals of Communication Theory U. Pillai & A. Patel

Modulation Property

If $f(t)$ is real,

$$F(\omega) = \int_{-\infty}^{+\infty} f(t)e^{-j\omega t}\,dt = \underbrace{\int_{-\infty}^{+\infty} f(t)\cos(\omega t)\,dt}_{R(\omega)} - j\underbrace{\int_{-\infty}^{+\infty} f(t)\sin(\omega t)\,dt}_{I(\omega)}$$

$$= R(\omega) - jI(\omega) = a - jb \tag{2.85}$$

$$|F(\omega)| = \sqrt{a^2 + b^2} = \sqrt{R^2(\omega) + I^2(\omega)} \tag{2.86}$$

$$R(\omega) = \int f(t)\cos(\omega t)\,dt = R(-\omega) \tag{2.87}$$

$$I(\omega) = \int f(t)\sin(\omega t)\,dt = -I(-\omega) \tag{2.88}$$

$$|F(\omega)|^2 = |F(-\omega)|^2 \tag{2.89}$$

If $f(t)$ is real, $|F(\omega)|$ is an even function.

Frequency Division Multiplexing (FDM)

Frequency Division Multiplexing refers to accommodating multiple base-band signals as in Fig. 3.1a by shifting each signal in the frequency domain using modulation, where the modulating frequency is adjusted so that none of the frequency transforms overlap. Thus,

$$f(t) = \sum_{k=1}^{N} f_k(t)\cos(k-1)2B_0 t \tag{2.90}$$

represents an FDM baseband signal. Here B_0 represents their common one-sided bandwidth.

Frequency Division Multiplexing (FDM)

If signals $f_1(t)$, $f_2(t)$, $f_3(t)$ have fourier transforms $F_1(\omega)$, $F_2(\omega)$ and $F_3(\omega)$ respectively;

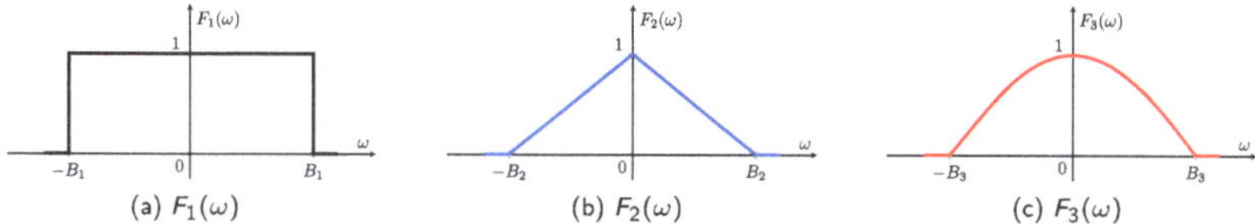

(a) $F_1(\omega)$ (b) $F_2(\omega)$ (c) $F_3(\omega)$

Figure 2.29: Individual Fourier Transform of three signals

Then after performing Frequency Division Multiplexing, the spectrum would look like:

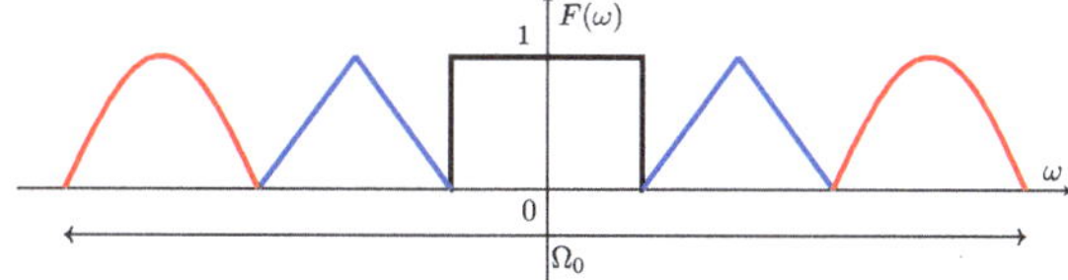

Figure 2.30: Spectrum after FDM

Frequency Division Multiplexing (FDM)

Finally f(t), a modulated set of signals as in (2.90), can be further modulated to any other frequency as

$$
\begin{aligned}
g(t) &= f(t)\cos\omega_0 t \\
G(\omega) &= \frac{1}{2}[F(\omega+\omega_0)+F(\omega-\omega_0)]
\end{aligned}
\tag{2.91}
$$

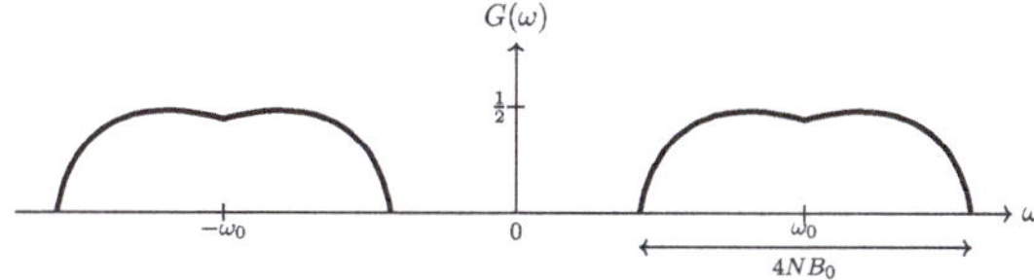

Figure 2.31: Fourier Transform representation of $g(t)$

Duality Property

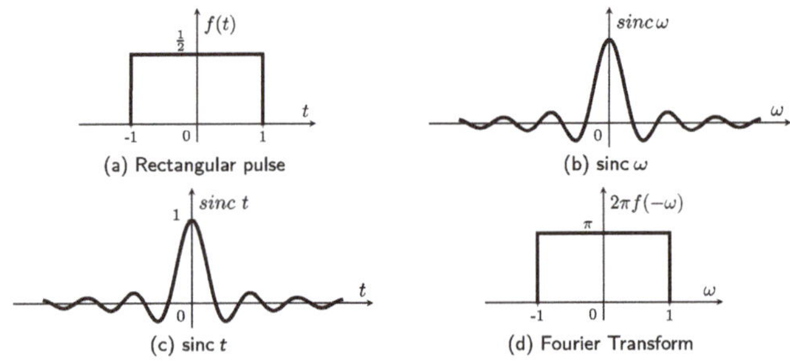

(a) Rectangular pulse (b) sinc ω (c) sinc t (d) Fourier Transform

Figure 2.32: Representation of Duality property of Fourier Transform

$$\begin{aligned} f(t) &\longleftrightarrow F(\omega) \\ F(t) &\longleftrightarrow ? \end{aligned} \tag{2.92}$$

$$F(\omega) - \int_{-\infty}^{\infty} F(t) e^{j\omega t}\, dt \tag{2.93}$$

$$f(t) = \frac{1}{2\pi} \int_{-\infty}^{\infty} F(\omega) e^{j\omega t} d\omega \tag{2.94}$$

Fourier Series Fourier Transforms Amplitude Modulation Angle Modulation Digital Modulation Techniques

Duality Property

Replacing ω with x,

$$2\pi f(t) = \int_{-\infty}^{\infty} F(x) e^{jxt}\, dx \tag{2.95}$$

Replacing t with $-\omega$,

$$2\pi f(-\omega) = \int_{-\infty}^{\infty} F(x) e^{-jx\omega}\, dx = \int_{-\infty}^{\infty} F(t) e^{-j\omega t}\, dt \tag{2.96}$$

$$\Rightarrow F(t) \longleftrightarrow 2\pi f(-\omega) \tag{2.97}$$

Duality Property,

$$\begin{aligned} f(t) &\longleftrightarrow F(\omega) \\ F(t) &\longleftrightarrow 2\pi f(-\omega) \end{aligned} \tag{2.98}$$

i.e., Any frequency transform shape is translated into the time domain, then the original time domain signal shape goes into frequency domain in a reversed manner.

Fundamentals of Communication Theory U. Pillai & A. Patel

Duality Property

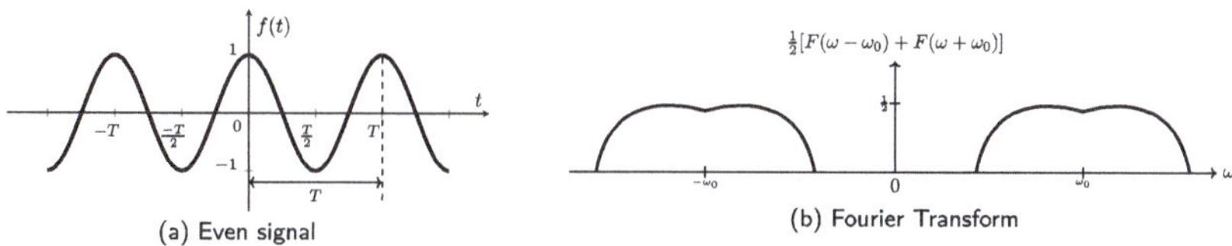

(a) Even signal

(b) Fourier Transform

Figure 2.33: Signal and its Fourier Transform

Even function: $f(t) = f(-t) \implies F(-\omega) = F(\omega)$, since

$$F(\omega) = \int_{-\infty}^{+\infty} f(t)e^{-j\omega t}\,dt = \int_{-\infty}^{+\infty} \underbrace{f(t)}_{\text{even}}\underbrace{\cos\omega t}_{\text{even}}\,dt - j\int_{-\infty}^{+\infty} \underbrace{f(t)}_{\text{even}}\underbrace{\sin\omega t}_{\text{odd}}$$
$$= 2\int_{0}^{\infty} f(t)\cos\omega t\,dt = F(-\omega). \tag{2.99}$$

Odd function: $f(t) = -f(t)$, then proceeding as above

$$F(\omega) = 2\int_{0}^{\infty} f(t)\sin\omega t\,dt = -F(-\omega) \tag{2.100}$$

Impulse Function

$$\text{Impulse or Delta function}, \delta(t) = 0, \quad t \neq 0 \tag{2.101}$$

$$\int_{-\infty}^{\infty} \delta(t)\,dt = 1 \tag{2.102}$$

$$F\{\delta(t)\} = \int_{-\infty}^{\infty} \delta(t)e^{-j\omega t}\,dt = 1 \tag{2.103}$$

Also,

$$\int_{-\infty}^{\infty} f(t_0 - \tau)\delta(\tau)\,d\tau = \int_{-\infty}^{\infty} f(\tau)\delta(t_0 - \tau)\,d\tau = f(t_0) \tag{2.104}$$

Fourier Series | **Fourier Transforms** | Amplitude Modulation | Angle Modulation | Digital Modulation Techniques

Impulse Function

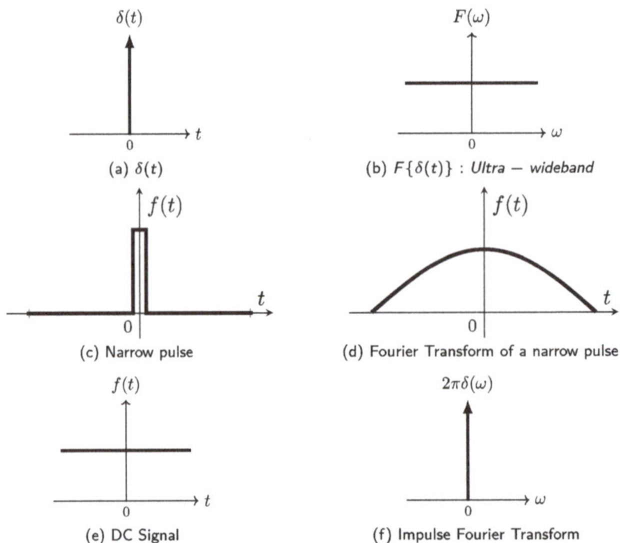

(a) $\delta(t)$ (b) $F\{\delta(t)\}$: *Ultra – wideband*

(c) Narrow pulse (d) Fourier Transform of a narrow pulse

(e) DC Signal (f) Impulse Fourier Transform

Figure 2.34: Signal and its Fourier Transform

Fundamentals of Communication Theory U. Pillai & A. Patel

Fourier Series | **Fourier Transforms** | Amplitude Modulation | Angle Modulation | Digital Modulation Techniques

Impulse Function

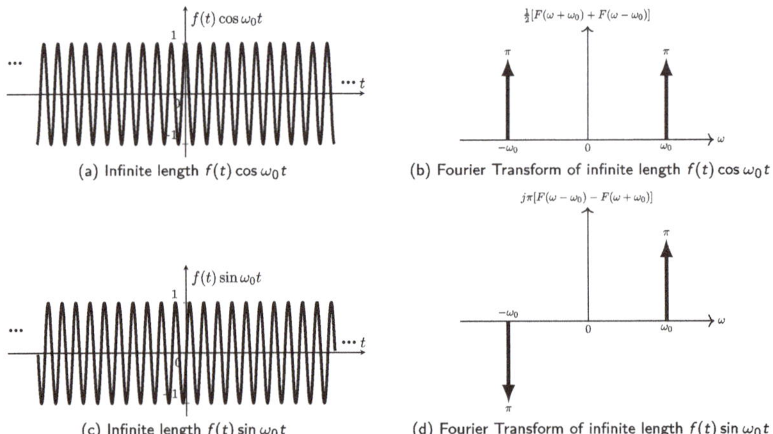

(a) Infinite length $f(t)\cos\omega_0 t$ (b) Fourier Transform of infinite length $f(t)\cos\omega_0 t$

(c) Infinite length $f(t)\sin\omega_0 t$ (d) Fourier Transform of infinite length $f(t)\sin\omega_0 t$

Figure 2.35: Signal and its Fourier Transform

Fundamentals of Communication Theory U. Pillai & A. Patel

Impulse Function

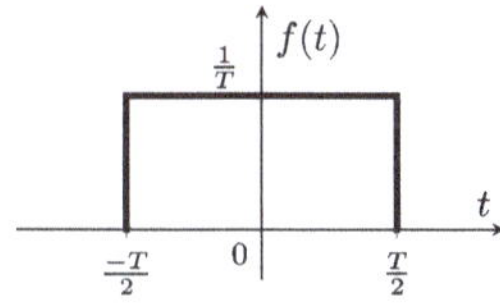

Figure 2.36: f(t)

Define $f_1(t)$ such that, $f_1(t) = f(t)\cos\omega_0 t$. Therefore,

$$FT\{f_1(t)\} = F_1(\omega) = \pi[F(\omega - \omega_0) + F(\omega + \omega_0)] \tag{2.105}$$

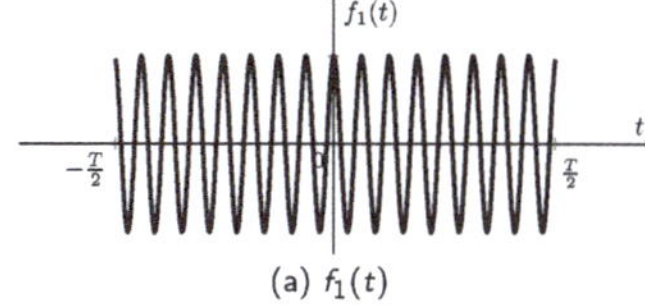

(a) $f_1(t)$

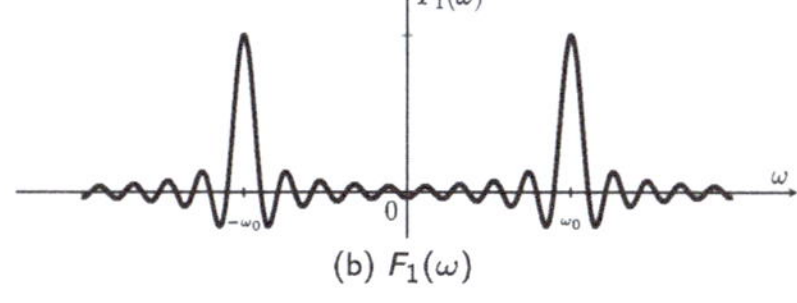

(b) $F_1(\omega)$

Figure 2.37: Signal and its Fourier Transform

Impulse Function

Define $p(t)$ such that, $p(t) = f(t)\sin\omega_0 t$. Therefore,

$$FT\{p(t)\} = P(\omega) = j\pi[F(\omega - \omega_0) - F(\omega + \omega_0)] \tag{2.106}$$

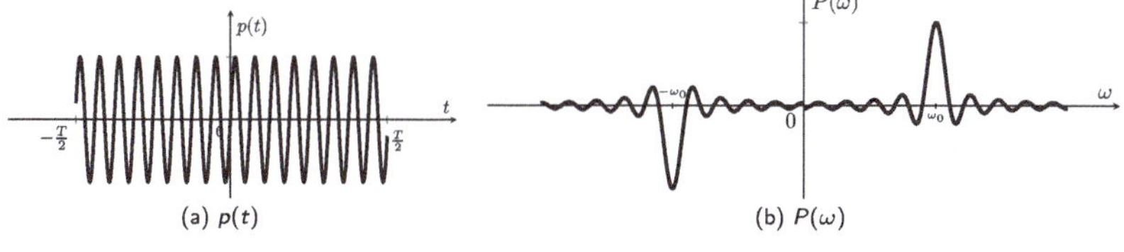

(a) $p(t)$ (b) $P(\omega)$

Figure 2.38: Signal and its Fourier Transform

Impulse Function

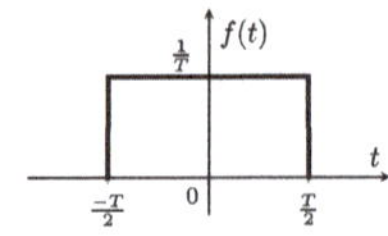

Figure 2.39: $f(t)$

$$\implies \mathcal{F}\{f(t)\} = \text{sinc}(\omega T/2)$$

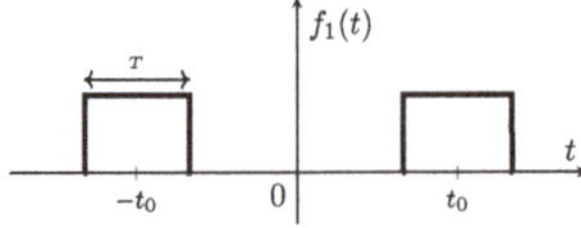

Figure 2.40: $f_1(t)$

$$\implies \mathcal{F}\{f_1(t)\} = 2\,\text{sinc}(\omega T/2)\cos\omega t_0$$

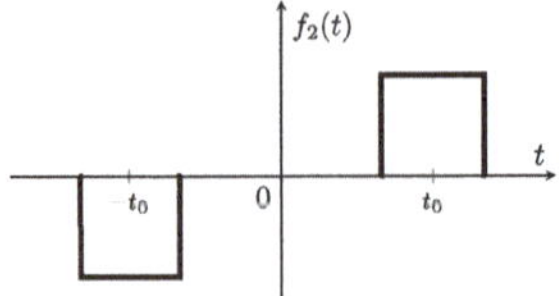

Figure 2.41: $f_2(t)$

$$\implies \mathcal{F}\{f_2(t)\} = 2j\,\text{sinc}(\omega T/2)\sin\omega t_0$$

Convolution Property

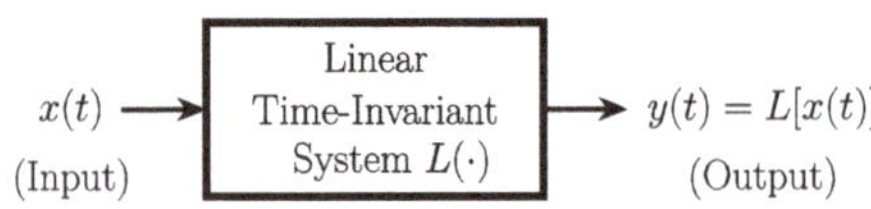

Figure 2.42: LTI System

System is linear

$$\text{If } y_1(t) = L[x_1(t)] \text{ and } y_2(t) = L[x_2(t)], \text{then}$$
$$L[a_1x_1(t) + a_1x_1(t)] = a_1L[x_1(t)] + a_2L[x_2(t)] = a_1\,y_1(t) + a_2\,y_2(t) \tag{2.107}$$

System is Time-Invariant

$$\text{If } L[x(t)] = y(t), \text{then } L[x(t-t_0)] = y(t-t_0) \tag{2.108}$$

To combine the properties in (2.107) and (2.108), write using (2.106)

$$x(t) = \int_{-\infty}^{\infty} x(\tau)\delta(t-\tau)\,d\tau \tag{2.109}$$

Convolution Property

and let

$$h(t) = L[\delta(t)] \tag{2.110}$$

represent the impulse response of the system above. Then

$$L[\delta(t-\tau)] = h(t-\tau) \tag{2.111}$$

Applying (2.107) in (2.109), we get the output response $y(t)$ to $x(t)$ to be

$$\begin{aligned} y(t) = L[x(t)] &= L\left[\int_{-\infty}^{\infty} x(\tau)\delta(t-\tau)\right] d\tau = \int_{-\infty}^{\infty} x(\tau) L\left[\delta(t-\tau)\right] d\tau \\ &= \int_{-\infty}^{\infty} x(\tau)\, h(t-\tau)\, d\tau = x(t) \circledast h(t), \end{aligned} \tag{2.112}$$

where the last operation represents the convolution of $x(t)$ with $h(t)$.

Convolution Property

Thus convolution of $h(t)$ with $g(t)$ can be written as:

$$h(t) \circledast g(t) = \int_{-\infty}^{\infty} h(\tau) g(t-\tau) d\tau = f(t) \tag{2.113}$$

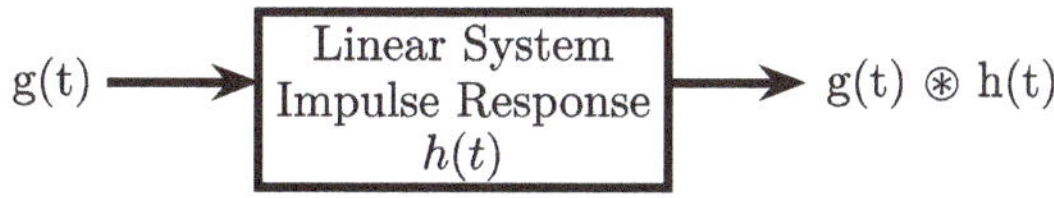

Figure 2.43: Linear System with impulse response $h(t)$

Example: Convolution of $g(t)$ with $h(t)$

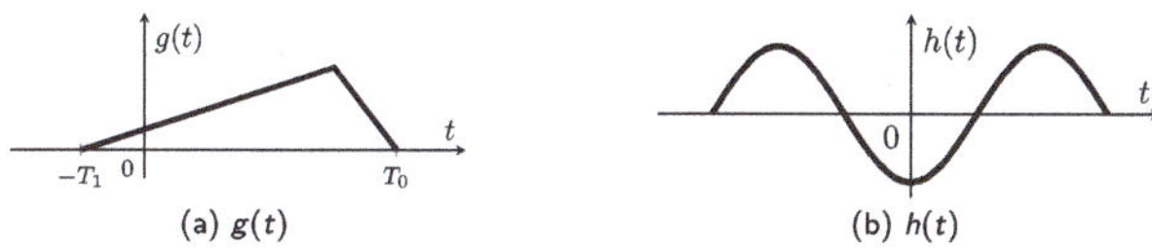

(a) $g(t)$ (b) $h(t)$

Figure 2.44: Convolution of $g(t)$ with $h(t)$

Convolution Property

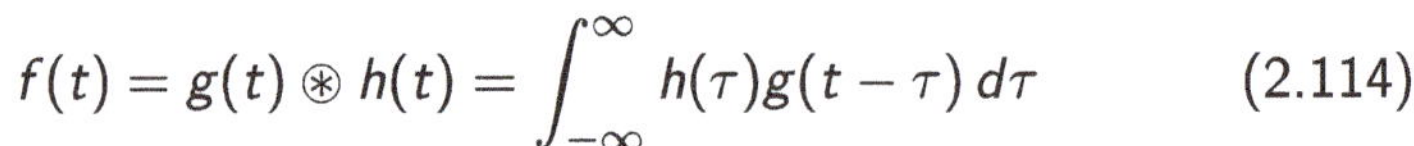

$$f(t) = g(t) \circledast h(t) = \int_{-\infty}^{\infty} h(\tau) g(t-\tau)\, d\tau \qquad (2.114)$$

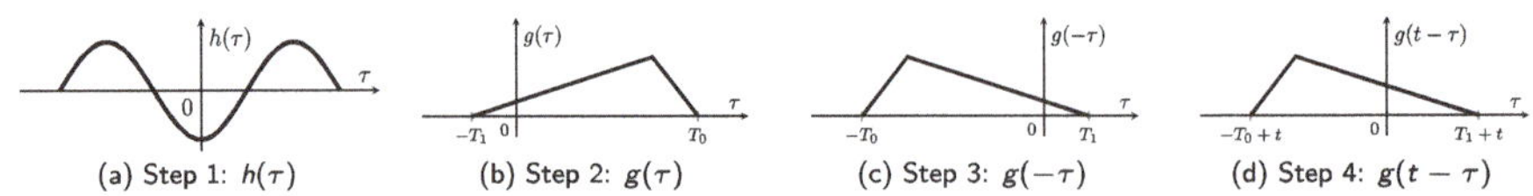

(a) Step 1: $h(\tau)$ (b) Step 2: $g(\tau)$ (c) Step 3: $g(-\tau)$ (d) Step 4: $g(t-\tau)$

Figure 2.45: Steps involved to calculate $f(t)$ which is convolution of $g(t)$ with $h(t)$

Step 5 is to multiply (a) and (d) and do the integration as in (2.114) for a given t. Repeat the above steps for every t. Then $f(t)$ can be represented as:

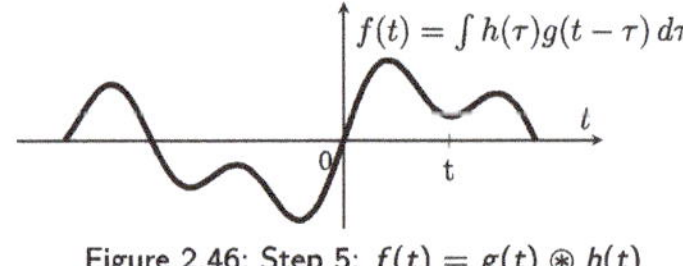

Figure 2.46: Step 5: $f(t) = g(t) \circledast h(t)$

Convolution Property

$$f(t) = f_1(t) \circledast f_2(t) = \int_{-\infty}^{\infty} f_1(\tau) f_2(t-\tau)\, d\tau \qquad (2.115)$$

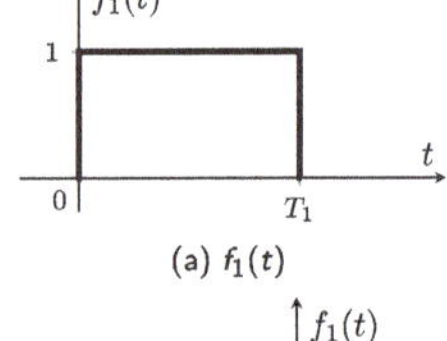

(a) $f_1(t)$

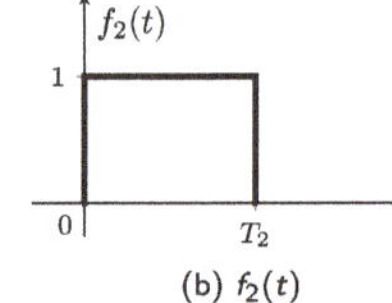

(b) $f_2(t)$

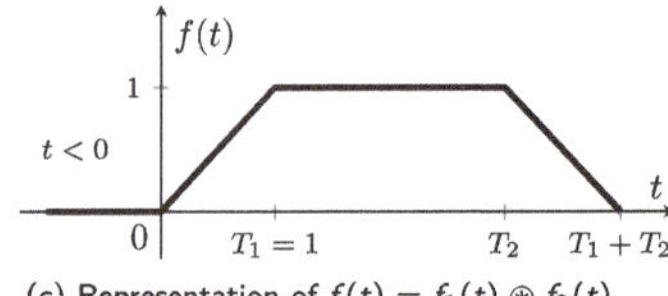

(c) Representation of $f(t) = f_1(t) \circledast f_2(t)$

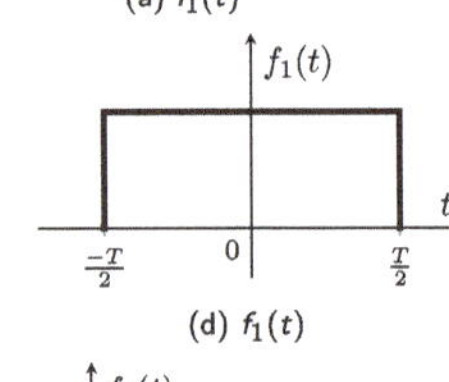

(d) $f_1(t)$

(e) $f_2(t)$

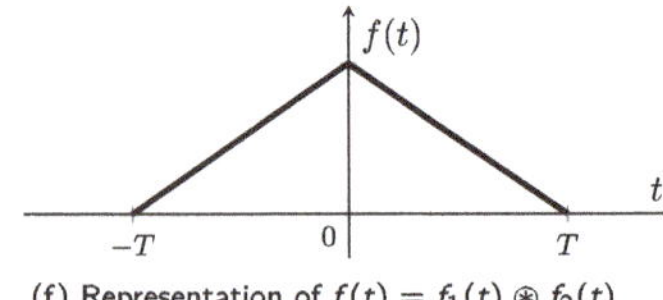

(f) Representation of $f(t) = f_1(t) \circledast f_2(t)$

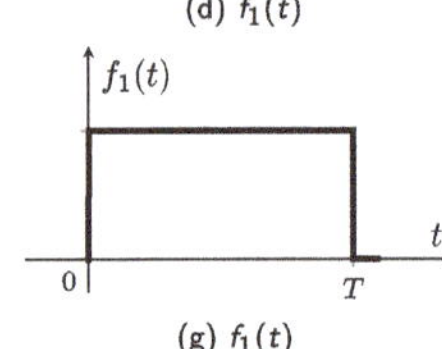

(g) $f_1(t)$

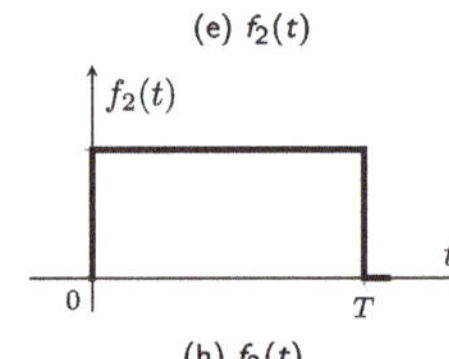

(h) $f_2(t)$

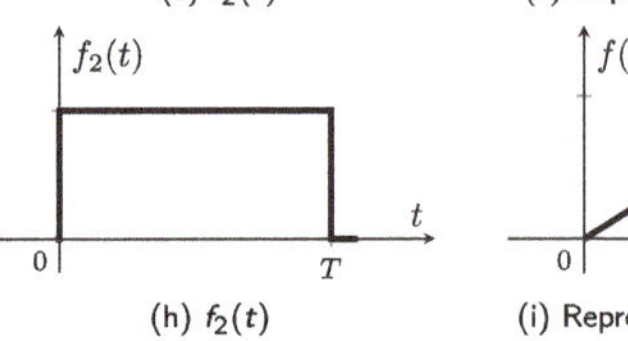

(i) Representation of $f(t) = f_1(t) \circledast f_2(t)$

Figure 2.47: Convolution of $f_1(t)$ with $f_2(t)$

Convolution Property

Convolution in time domain results in multiplication in the frequency domain. To see this, let

$$h(t) \longleftrightarrow H(\omega), g(t) \longleftrightarrow G(\omega) \tag{2.116}$$

$$f(t) = h(t) \circledast g(t) = \int_{-\infty}^{\infty} h(\tau)g(t-\tau)\, d\tau \tag{2.117}$$

$$F(\omega) = \int_{-\infty}^{\infty}\int_{-\infty}^{\infty} h(\tau)g(t-\tau)e^{-j\omega t}\, dt\, d\tau \tag{2.118}$$

Replacing $t-\tau$ with x. Therefore $t = x+\tau$

$$= \int_{-\infty}^{\infty}\int_{-\infty}^{\infty} h(\tau)g(x)e^{-j\omega(x+\tau)}\, d\tau\, dx \tag{2.119}$$

$$= \underbrace{\int_{-\infty}^{\infty} h(\tau)e^{-j\omega\tau}\, d\tau}_{H(\omega)} \underbrace{\int_{-\infty}^{\infty} g(x)e^{-j\omega x}\, dx}_{G(\omega)} \tag{2.120}$$

or

$$F(\omega) = H(\omega)G(\omega) \tag{2.121}$$

Fourier Series Fourier Transforms Amplitude Modulation Angle Modulation Digital Modulation Techniques

Convolution Property

$$f(t) = g(t) \circledast h(t) \tag{2.122}$$

$$f(t) = \int g(\tau)h(t-\tau)d\tau \tag{2.123}$$

$$F(\omega) = G(\omega)H(\omega) \tag{2.124}$$

$$G(\omega) = \mathcal{F}\{g(t)\} \tag{2.125}$$

$$H(\omega) = \mathcal{F}\{h(t)\} \tag{2.126}$$

$$\mathcal{F}\{h(t) \circledast g(t)\} = H(\omega)\cdot G(\omega) \tag{2.127}$$

Fundamentals of Communication Theory U. Pillai & A. Patel

Convolution Property

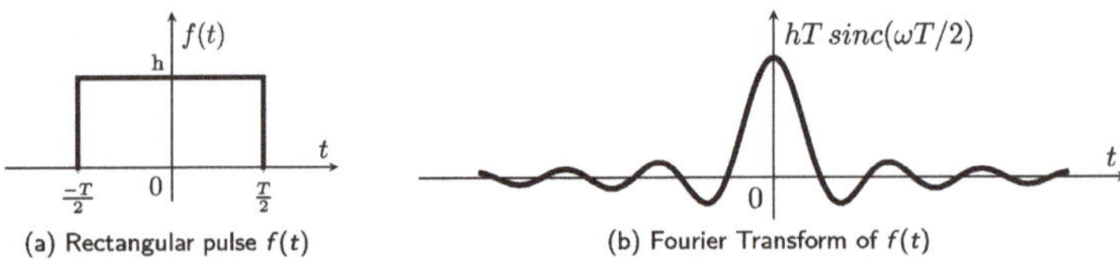

(a) Rectangular pulse $f(t)$ (b) Fourier Transform of $f(t)$

Figure 2.48: Signal and its Fourier Transform

$$p(t) = f(t) \circledast f(t) \quad \text{, is shown below} \tag{2.128}$$

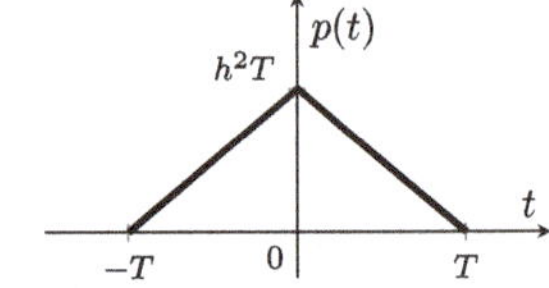

Figure 2.49: Representation of $p(t) = f(t) \circledast f(t)$

$$\mathcal{F}\{p(t)\} = P(\omega) = h^2\, T^2 \cdot sinc^2\,(\omega T/2) \tag{2.129}$$

Convolution Property

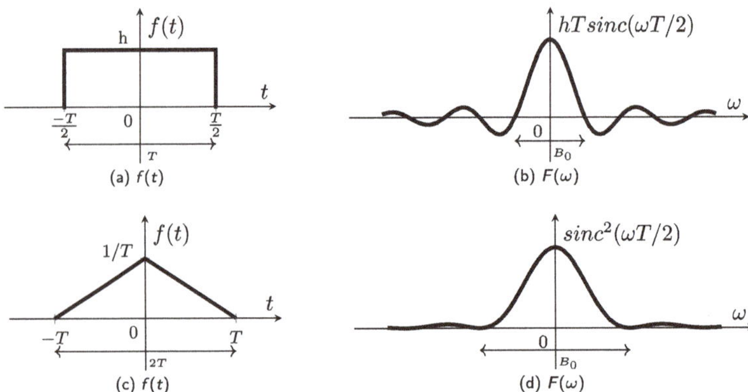

(a) $f(t)$ (b) $F(\omega)$

(c) $f(t)$ (d) $F(\omega)$

Figure 2.50: Signal and its Fourier Transform

Pulse shape plays a major role in determining the bandwidth.

Convolution Property

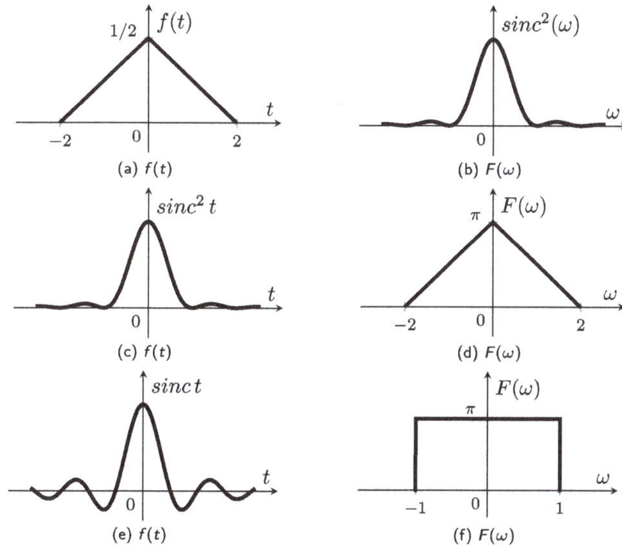

Figure 2.51: Signal and its Fourier Transform

Fourier Series Fourier Transforms Amplitude Modulation Angle Modulation Digital Modulation Techniques

Multiplication in Time

$$h(t) \underset{\substack{\uparrow \\ convolution}}{\circledast} g(t) \Leftrightarrow H(\omega)G(\omega) \tag{2.130}$$

$$f(t) = h(t) \cdot g(t) \longleftrightarrow ? \tag{2.131}$$

$$F(\omega) = \int_{-\infty}^{+\infty} f(t)e^{-j\omega t}\, dt = \int h(t) \cdot g(t)e^{-j\omega t}\, dt \tag{2.132}$$

$$F(\omega) = \frac{1}{2\pi} G(\omega) \circledast H(\omega) \tag{2.133}$$

But

$$g(t) = \frac{1}{2\pi} \int G(x)e^{jxt}\, dx \tag{2.134}$$

Hence

$$F(\omega) = \frac{1}{2\pi} \int_{-\infty}^{\infty} h(t) \int_{-\infty}^{\infty} G(x)e^{jxt}\, e^{-j\omega t}\, dx\, dt \tag{2.135}$$

Multiplication in Time

After changing the order of integration,

$$\begin{aligned} F(\omega) &= \frac{1}{2\pi}\int_{-\infty}^{+\infty} G(x)\left(\int_{-\infty}^{+\infty} h(t)e^{-j(\omega-x)t}\,dt\right)dx \\ &= \frac{1}{2\pi}\int_{-\infty}^{+\infty} G(x)H(\omega-x)\,dx = \frac{1}{2\pi}F(\omega)\circledast H(\omega) \end{aligned} \tag{2.136}$$

$$g(t) \underset{\substack{\uparrow \\ \text{convolve}}}{\circledast} h(t) \longleftrightarrow G(\omega) \underset{\substack{\uparrow \\ \text{multiply}}}{\cdot} H(\omega) \tag{2.137}$$

$$g(t) \underset{\substack{\uparrow \\ \text{multiply}}}{\cdot} h(t) \longleftrightarrow \frac{1}{2\pi}G(\omega) \underset{\substack{\uparrow \\ \text{convolve}}}{\circledast} H(\omega) \tag{2.138}$$

Thus convolution in the time domain results in multiplication in the frequency domain and multiplication in the time domain results in convolution in the frequency domain, exhibiting duality.

Fourier Transform Properties

$$f(t) = \int_{-\infty}^{\infty} F(\omega)e^{j\omega t}d\omega \tag{2.139}$$

$$f'(t) = \frac{df(t)}{dt} = \int_{-\infty}^{\infty} j\omega F(\omega)e^{j\omega t}d\omega \tag{2.140}$$

$$f'(t) \longleftrightarrow j\omega F(\omega) \tag{2.141}$$

and

$$f^{(n)}(t) \longleftrightarrow (j\omega)^n F(\omega) \tag{2.142}$$

Fourier Transform Properties

Table 2.1: Fourier Transform Pairs

Signal	*Fourier Transform*
$f(t)$	$F(\omega)$
$a_1 f_1(t) + a_2 f_2(t)$	$a_1 F_1(\omega) + a_2 F_2(\omega)$
$f(t - t_0)$	$e^{-j\omega t_0} F(\omega)$
$f(t)e^{j\omega_0 t}$	$F(\omega - \omega_0)$
$f(t)e^{-j\omega_0 t}$	$F(\omega + \omega_0)$
$f(t)\cos(\omega_0 t)$	$\frac{1}{2}[F(\omega + \omega_0) + F(\omega - \omega_0)]$
$f(t)\sin(\omega_0 t)$	$\frac{j}{2}[F(\omega + \omega_0) - F(\omega - \omega_0)]$
$f(at)$	$\frac{1}{\lvert a \rvert} F\left(\frac{\omega}{a}\right)$

Fourier Series Fourier Transforms Amplitude Modulation Angle Modulation Digital Modulation Techniques

Fourier Transform Properties

Table 2.2: Fourier Transform Pairs

Signal	*Fourier Transform*
$F(t)$	$2\pi f^*(-\omega)$
$tf(t)$	$jF'(\omega)$
$f'(t)$	$j\omega F(\omega)$
$\hat{f}(t) = f(t) \circledast \frac{1}{\pi t}$	$j \operatorname{sgn} \omega \cdot F(\omega)$
$\delta(t)$	1
1	$2\pi\delta(\omega)$
$f(t) \circledast h(t)$	$F(\omega) \cdot H(\omega)$
$f(t) \cdot h(t)$	$\frac{1}{2\pi} F(\omega) \circledast H(\omega)$

Fundamentals of Communication Theory U. Pillai & A. Patel

Sign function

Let sgn(t) represent the sign function

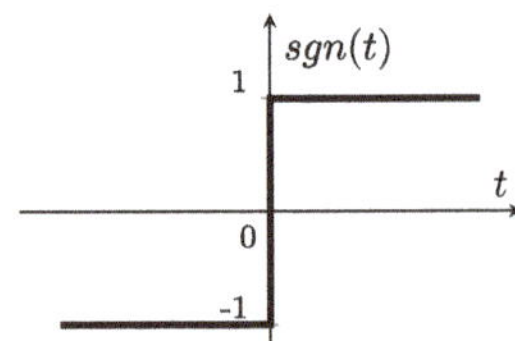

Figure 2.52: Sign function

Thus

$$f(t) = \text{sgn}(t) = \begin{cases} +1, & t \geq 0 \\ -1, & t < 0 \end{cases} \tag{2.143}$$

Then

$$f'(t) = 2\delta(t) \tag{2.144}$$

Fourier Series Fourier Transforms Amplitude Modulation Angle Modulation Digital Modulation Techniques

Sign function & Hilbert Transform

Using (2.141),

$$j\omega F(\omega) = \mathcal{F}\{f'(t)\} = 2\mathcal{F}\{\delta(t)\} = 2 \tag{2.145}$$

$$F(\omega) = \mathcal{F}\{\text{sgn}(t)\} = \frac{2}{j\omega} \tag{2.146}$$

Thus

$$f(t) = sgn(t) \longleftrightarrow F(\omega) = \frac{2}{j\omega} \tag{2.147}$$

and using duality in (2.98),

$$F(t) = \frac{2}{jt} \longleftrightarrow 2\pi f(-\omega) = 2\pi sgn(-\omega) \tag{2.148}$$

Or,

$$\frac{1}{jt} \longleftrightarrow -\pi sgn(\omega) \tag{2.149}$$

Fundamentals of Communication Theory U. Pillai & A. Patel

Sign function & Hilbert Transform

$$g(t) = \underbrace{\frac{1}{\pi t}}_{\text{real}} \Longleftrightarrow G(\omega) = -j\,\text{sgn}(\omega) \tag{2.150}$$

Define

$$\hat{f}(t) \Longleftrightarrow F(\omega) \cdot G(\omega) \tag{2.151}$$

$G(\omega) = -j\,\text{sgn}(\omega)$ in (2.150) is known as the Hilbert Transform operator.

$$\underbrace{\hat{F}(\omega)}_{\hat{f}(t)} = \underbrace{F(\omega) \cdot G(\omega)}_{f(t)\circledast g(t)} = -jF(\omega)\,\text{sgn}(\omega) \tag{2.152}$$

Then

$$\hat{f}(t) = f(t) \circledast g(t) = \int_{-\infty}^{\infty} f(\tau) g(t-\tau)\, d\tau = \frac{1}{\pi} \int_{-\infty}^{\infty} \frac{f(\tau)}{t-\tau}\, d\tau \tag{2.153}$$

$\hat{f}(t)$ is the Hilbert Transform of $f(t)$ in the time domain.

$f(t)$ → [Fourier Transform] → $F(\omega)$ → [$G(\omega) = -j\,\text{sgn}(\omega)$] → $\hat{F}(\omega)$ → [Inverse F.T.] → Hilbert Transform of $f(t)$: $\hat{f}(t) = f(t) \circledast g(t)$

Figure 2.53: Hilbert Transform

Hilbert Transform

Hilbert Transform is a quadrature phase shifter (phase shift by $-\pi/2$) such that the corresponding operation in time domain is implemented through a real filter

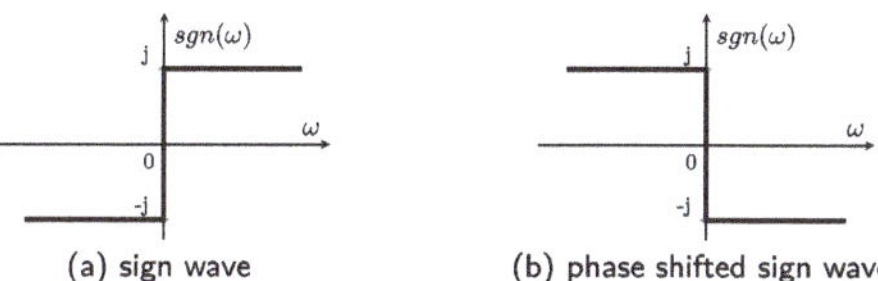

(a) sign wave (b) phase shifted sign wave

Figure 2.54: Hilbert Transform as a phase shifter

$$\begin{aligned} g(t) &= \frac{1}{\pi t} \longleftrightarrow G(\omega) = -j\ \text{sgn}(\omega) \\ f(t) &\longleftrightarrow F(\omega) \\ \hat{f}(t) &\longleftrightarrow \underbrace{F(\omega)G(\omega)}_{\hat{F}(\omega)} = -j\,F(\omega)\ \text{sgn}(\omega), \end{aligned} \tag{2.154}$$

where

$$\hat{f}(t) = \frac{1}{\pi} \int_{-\infty}^{+\infty} \frac{f(\tau)}{t-\tau}\, d\tau$$

Hilbert Transform

$$\underbrace{f(t)}_{\text{real}} \longleftrightarrow F(\omega) = F(-\omega) \tag{2.155}$$

$$f(t)\cos\omega_0 t \longleftrightarrow \frac{1}{2}\left[F(\omega-\omega_0) + F(\omega+\omega_0)\right] \tag{2.156}$$

$$\begin{aligned}\hat{f}(t)\sin\omega_0 t \longleftrightarrow & \frac{1}{2j}\left[\hat{F}(\omega-\omega_0) - \hat{F}(\omega+\omega_0)\right] \\ = & \frac{1}{2}[F(\omega+\omega_0)\,\mathrm{sgn}(\omega+\omega_0) - F(\omega-\omega_0)\,\mathrm{sgn}(\omega-\omega_0)]\end{aligned} \tag{2.157}$$

Fourier Series Fourier Transforms Amplitude Modulation Angle Modulation Digital Modulation Techniques

Hilbert Transform

Adding equations (2.156) and (2.157) and using (2.152), we get

$$\begin{aligned}g_1(t) &= f(t)\cos\omega_0 t + \hat{f}(t)\sin\omega_0 t \longleftrightarrow \\ G_1(\omega) &= \frac{1}{2}\left\{F(\omega+\omega_0)[1+\mathrm{sgn}(\omega+\omega_0)] + F(\omega-\omega_0)[1-\mathrm{sgn}(\omega-\omega_0)]\right\}\end{aligned} \tag{2.158}$$

Subtracting equations (2.156) and (2.157), we get

$$\begin{aligned}g_2(t) &= f(t)\cos\omega_0 t - \hat{f}(t)\sin\omega_0 t \longleftrightarrow \\ G_2(\omega) &= \frac{1}{2}[F(\omega+\omega_0)(1-\mathrm{sgn}(\omega-\omega_0)) - F(\omega-\omega_0)(1+\mathrm{sgn}(\omega+\omega_0))]\end{aligned} \tag{2.159}$$

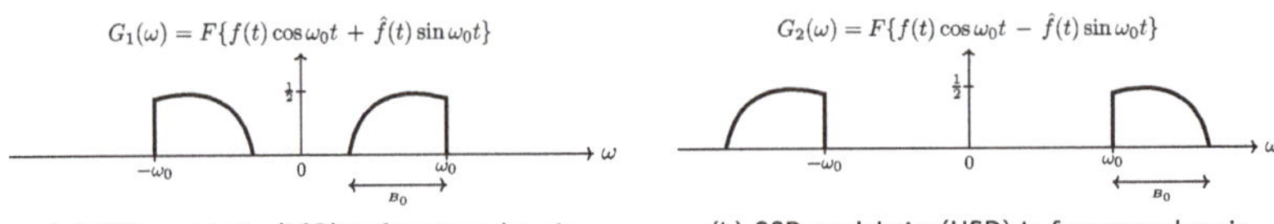

(a) SSB modulation(LSB) in frequency domain (b) SSB modulation(USB) in frequency domain

Figure 2.55: SSB modulated signal in frequency domain

Amplitude Modulation

$g_1(t) = f(t)\cos\omega_0 t$ represents double sideband modulation (DSB) and $g_2(t) = [1 + \lambda f(t)]\cos\omega_0 t$ represents Amplitude Modulation (AM).

(a) Block Diagram for AM Modulation (b) AM signal in frequency domain

Figure 2.56: Amplitude Modulation

Block diagram to generate a Single Side Band (SSB) signal

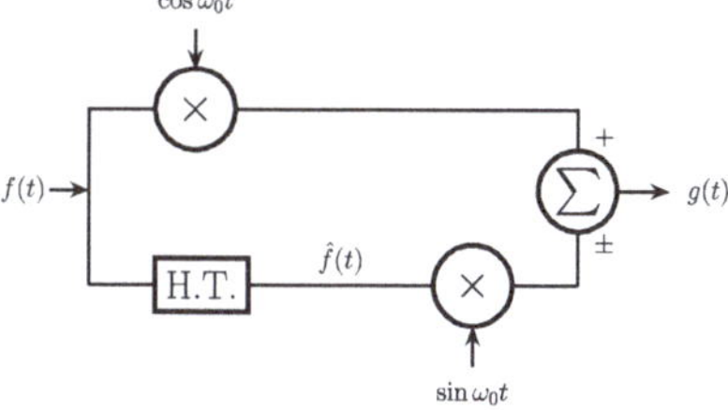

Fourier Series Fourier Transforms Amplitude Modulation Angle Modulation Digital Modulation Techniques

Single Side-Band (SSB) Modulation

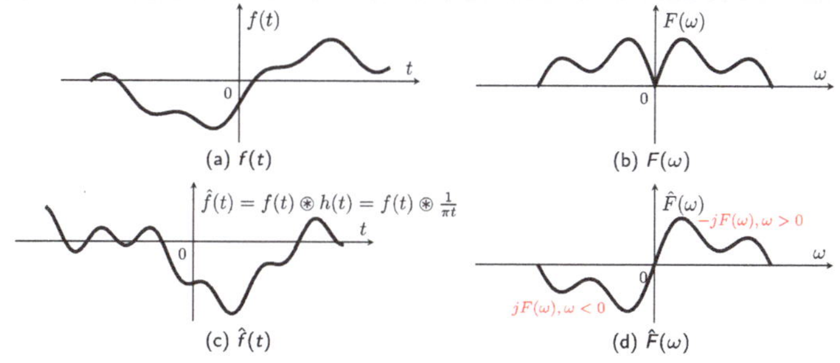

(a) $f(t)$ (b) $F(\omega)$

(c) $\hat{f}(t)$ (d) $\hat{F}(\omega)$

Figure 2.58: Signal and its Fourier Transform

$$g(t) = f(t)\cos\omega_0 t + \hat{f}(t)\sin\omega_0 t \quad \text{, where} \qquad (2.160)$$
$$\hat{f}(t) = f(t) \circledast \frac{1}{\pi t}$$

$f(t)$ is the in-phase signal and $\hat{f}(t)$ is the quadrature phase signal. Hilbert Transform helps to reduce the modulated signal bandwidth by a factor of 2.

Hilbert Transform

Let $x(t) = f(t)\cos\omega_0 t$ and $y(t) = \hat{f}(t)\sin\omega_0 t$. Therfore,

$$x(t) = f(t)\cos\omega_0 t = f(t)\left(\frac{e^{-j\omega_0 t} + e^{j\omega_0 t}}{2}\right) \tag{2.161}$$

$$X(\omega) = \frac{1}{2}\left(F(\omega+\omega_0) + F(\omega-\omega_0)\right) \tag{2.162}$$

$$\int_{-\infty}^{\infty} f(t)e^{-j\omega_0 t}e^{-j\omega t}\,dt = \int_{-\infty}^{\infty} f(t)e^{-j(\omega+\omega_0)t}\,dt \tag{2.163}$$

$$y(t) = \hat{f}(t)\sin\omega_0 t = \hat{f}(t)\left(\frac{e^{j\omega_0 t} - e^{-j\omega_0 t}}{2j}\right) \tag{2.164}$$

$$\hat{F}(\omega) = H(\omega)F(\omega) = -j\,\mathrm{sgn}(\omega)F(\omega) \tag{2.165}$$

$$\text{Therefore, } Y(\omega) = \frac{1}{2j}\left(\hat{F}(\omega-\omega_0) - \hat{F}(\omega+\omega_0)\right) \tag{2.166}$$

Hilbert Transform

$$\begin{aligned} Y(\omega) &= \frac{1}{2j}\left[-j\,\mathrm{sgn}(\omega-\omega_0)F(\omega-\omega_0) - (-j)\,\mathrm{sgn}(\omega+\omega_0)F(\omega+\omega_0)\right] \\ &= \frac{1}{2}\left[F(\omega+\omega_0)\,\mathrm{sgn}(\omega+\omega_0) - F(\omega-\omega_0)\,\mathrm{sgn}(\omega-\omega_0)\right] \end{aligned} \tag{2.167}$$

$$X(\omega) = \frac{1}{2}\left[F(\omega+\omega_0) + F(\omega-\omega_0)\right] \tag{2.168}$$

$$\begin{aligned} G(\omega) &= X(\omega) + Y(\omega) \\ &= \frac{1}{2}\left[\underbrace{(1+\mathrm{sgn}(\omega+\omega_0))}_{2U(\omega+\omega_0)}F(\omega+\omega_0) + \underbrace{(1-\mathrm{sgn}(\omega-\omega_0))}_{2U(-\omega+\omega_0)}F(\omega-\omega_0)\right] \\ &= F(\omega+\omega_0)U(\omega+\omega_0) + F(\omega-\omega_0)U(-\omega+\omega_0) \end{aligned} \tag{2.169}$$

Hilbert Transform

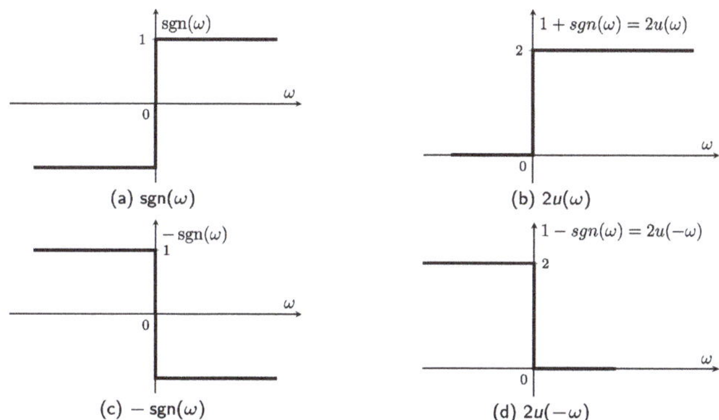

(a) $\mathrm{sgn}(\omega)$ (b) $2u(\omega)$ (c) $-\mathrm{sgn}(\omega)$ (d) $2u(-\omega)$

Figure 2.59: Signal and its Fourier Transform

$$\begin{aligned} 1 + \mathrm{sgn}(\omega + \omega_0) &= 2U(\omega + \omega_0) \\ 1 - \mathrm{sgn}(\omega - \omega_0) &= 2U(-\omega + \omega_0) \end{aligned} \tag{2.170}$$

Hilbert Transform

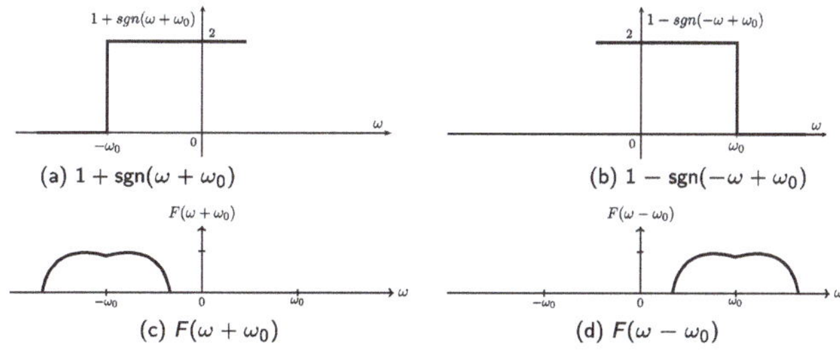

(a) $1 + \mathrm{sgn}(\omega + \omega_0)$ (b) $1 - \mathrm{sgn}(-\omega + \omega_0)$ (c) $F(\omega + \omega_0)$ (d) $F(\omega - \omega_0)$

Figure 2.60: Generating SSB modulated signal

Multiplying signals in (a) and (b) with (c) and (d) respectively and adding them results in:

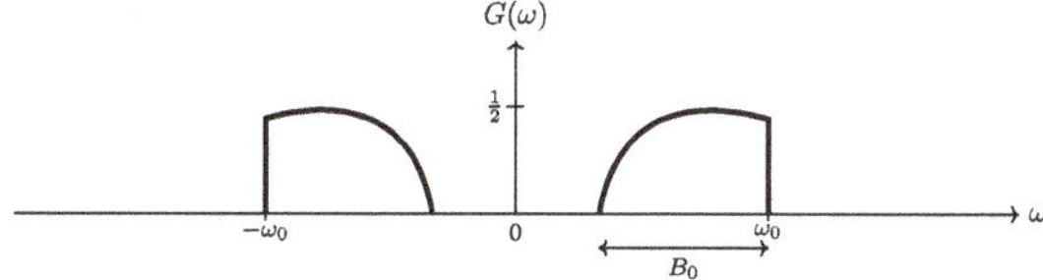

Figure 2.61: SSB modulation(LSB) in frequency domain

Hilbert Transform

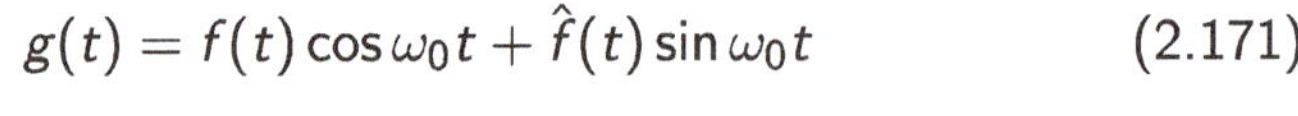

$$g(t) = f(t)\cos\omega_0 t + \hat{f}(t)\sin\omega_0 t \tag{2.171}$$

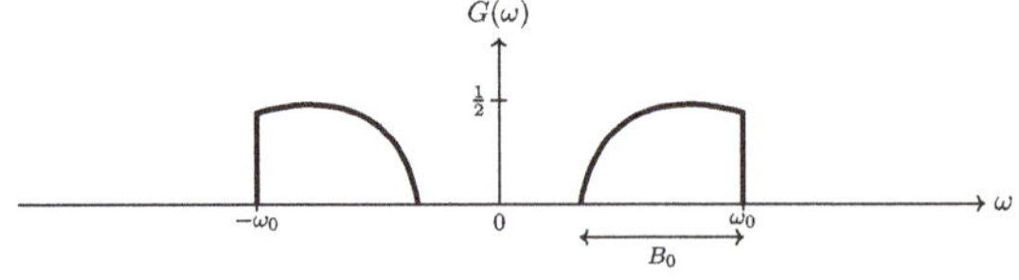

Figure 2.62: SSB modulation(LSB) in frequency domain

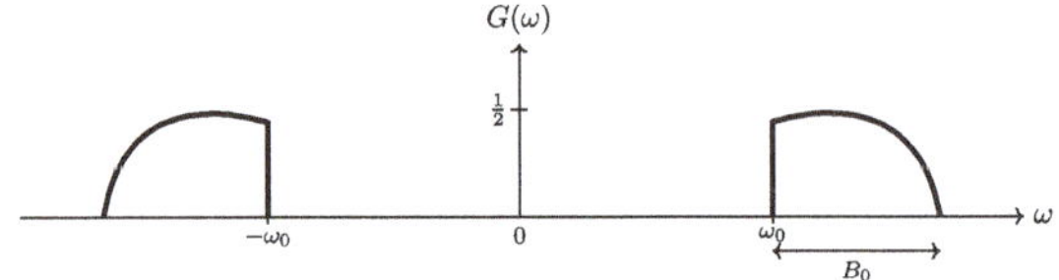

$$g_1(t) = f(t)\cos\omega_0 t - \hat{f}(t)\sin\omega_0 t \tag{2.172}$$

Figure 2.63: SSB modulation(USB) in frequency domain

Fourier Series Fourier Transforms Amplitude Modulation Angle Modulation Digital Modulation Techniques

Hilbert Transform (Alternate Approach)

How to find the Hilbert transforms of signals?
Consider

$$h(t) = \frac{1}{\pi}\log|t| \tag{2.173}$$

$$h'(t) = \begin{cases} \frac{1}{\pi}\cdot\frac{1}{t}, & t > 0 \\ \frac{1}{\pi}\cdot\frac{1}{-(-t)}, & t < 0 \end{cases}$$

$$h'(t) = \frac{1}{\pi t} \longleftrightarrow -j\,\mathrm{sgn}\,\omega \tag{2.174}$$

On taking Fourier Transform, we get

$$j\omega H(\omega) = -j\,\mathrm{sgn}(\omega) \tag{2.175}$$

or

$$H(\omega) = \frac{-\mathrm{sgn}(\omega)}{\omega} \tag{2.176}$$

Fundamentals of Communication Theory U. Pillai & A. Patel

Hilbert Transform (Alternate Approach)

$$\hat{f}(t) = f(t) \circledast \frac{1}{\pi t} \tag{2.177}$$

$$\hat{F}(\omega) = F(\omega) \cdot (-j\,\mathrm{sgn}(\omega)) = j\omega F(\omega) \underbrace{\left(\frac{-\mathrm{sgn}(\omega)}{\omega}\right)}_{H(\omega)} \tag{2.178}$$

From (2.173) and (2.175)

$$\hat{f}(t) = f'(t) \circledast h(t) \tag{2.179}$$

where

$$h(t) = \frac{1}{\pi} \log|t|$$

$$\begin{aligned}\hat{f}(t) &= f'(t) \circledast \frac{1}{\pi}\log|t| = f(t) \circledast \frac{1}{\pi t} = \frac{1}{\pi}\int_{-\infty}^{\infty} \frac{f(\tau)}{t-\tau}\,d\tau \\ &= \frac{1}{\pi}\int_{-\infty}^{\infty} f'(\tau)\log|t-\tau|\,d\tau\end{aligned} \tag{2.180}$$

Fourier Series Fourier Transforms Amplitude Modulation Angle Modulation Digital Modulation Techniques

Hilbert Transform (Alternate Approach)

Figure 2.64: Signal and its Hilbert Transform

$$f'(\tau) = \delta\left(\tau + \frac{T}{2}\right) - \delta\left(\tau - \frac{T}{2}\right) \tag{2.181}$$

$$\begin{aligned}\hat{f}(t) &= \frac{1}{\pi}\int_{-T/2}^{T/2} f'(\tau)\log|t-\tau|\,d\tau \\ &= \int_{-T/2}^{T/2}\left[\delta\left(\tau+\frac{T}{2}\right) - \delta\left(\tau-\frac{T}{2}\right)\right]\log|t-\tau|\,d\tau \\ &= \log\left|t+\frac{T}{2}\right| - \log\left|t-\frac{T}{2}\right| = \log\left|\frac{t+T/2}{t-T/2}\right|\end{aligned} \tag{2.182}$$

Note that sharp edges in $f(t)$ corresponds to peaks in $\hat{f}(t)$.

Fundamentals of Communication Theory U. Pillai & A. Patel

Nyquist's Theorem

$$\text{Bandlimited Signals: } F(\omega) = 0, |\omega| > B_0 \tag{2.183}$$

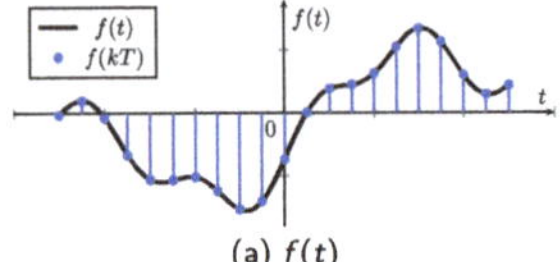

(a) $f(t)$

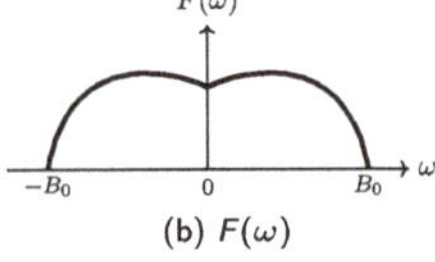

(b) $F(\omega)$

Figure 2.65: Bandlimited signals

Nyquist's Theorem: Any bandlimited signal can be reconstructed from its samples, provided the sampling rate $T \leq T_n$, where the Nyquist's rate T_N is given by:

$$T_N = \frac{2\pi}{2B_0} = \frac{\pi}{B_0} = \frac{\pi}{2\pi f_0} = \frac{1}{2f_0}$$

Fourier Series | Fourier Transforms | Amplitude Modulation | Angle Modulation | Digital Modulation Techniques

Nyquist's Theorem

Proof: $f(t)$ is bandlimited signal, then

$$F(\omega) = 0, \quad |\omega| > B_0 \tag{2.184}$$

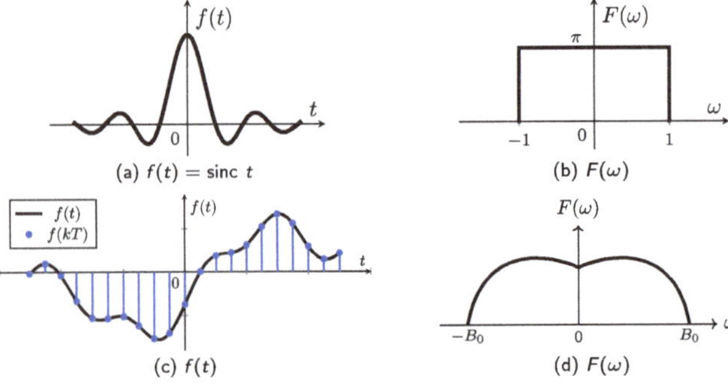

Figure 2.66: Bandlimited signals

Fundamentals of Communication Theory U. Pillai & A. Patel

Nyquist's Theorem

In Fig. 2.66d, extend the band-limited function $F(\omega)$ periodically to generate $F_p(\omega)$ that is periodic in the frequency domain. Then $F_p(\omega)$ has a Fourier Series representation in the frequency domain as shown in Fig. 2.67

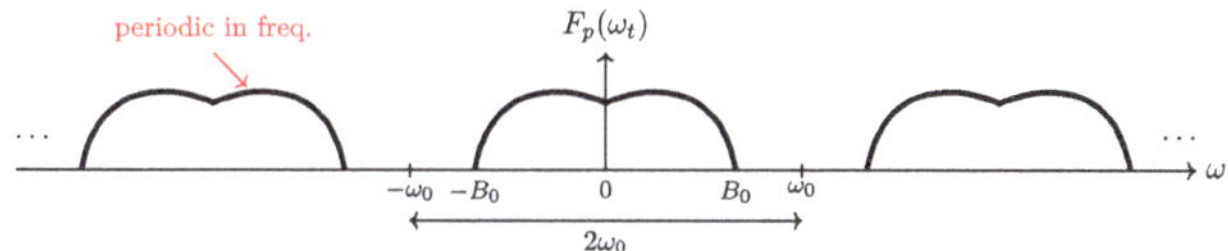

Figure 2.67: Bandlimited signal in frequency domain with period $2\omega_0$

$$f_p(t) = \sum_{k=-\infty}^{\infty} c_k e^{-jk\omega_0 t} \tag{2.185}$$

where

$$c_k = \frac{1}{T} \int_{-T/2}^{T/2} f(t) e^{jk\omega_0 t} \, dt \tag{2.186}$$

Sampling Theorem

Thus with $T = \frac{2\pi}{2\omega_0}$

$$F_p(\omega) = \sum_{k=-\infty}^{\infty} c_k e^{-jk\left(\frac{2\pi}{2\omega_0}\right)\omega_0} = \sum_{k=-\infty}^{\infty} c_k e^{-jkT\omega} \tag{2.187}$$

$$c_k = \frac{2\pi}{2\omega_0} \cdot \frac{1}{2\pi} \int_{-\omega_0}^{\omega_0} F(\omega) e^{-jkT\omega} \, d\omega = \frac{2\pi}{2\omega_0} \cdot f(kT) = T \cdot f(kT) \tag{2.188}$$

$$F_p(\omega) = T \sum_{k=-\infty}^{\infty} f(kT) e^{-jkT\omega} \tag{2.189}$$

Figure 2.68: $H(\omega)$

Observe that $F(\omega)$ can be reconstructed from $F_p(\omega)$ using $H(\omega)$ in Fig. 3.2 as follows

$$F(\omega) = F_p(\omega) H(\omega) \tag{2.190}$$

Sampling Theorem

$$F(\omega) = F_p(\omega)H(\omega) = T \sum_{k=-\infty}^{\infty} f(kT)e^{-jkT\omega} H(\omega)$$
$$= T \sum_{k=-\infty}^{\infty} f(kT)H(\omega)e^{-jkT\omega} \tag{2.191}$$

$$h(t) \longleftrightarrow H(\omega)$$
$$h(t-kT) \longleftrightarrow H(\omega)e^{-j\omega kT} \tag{2.192}$$

Taking Inverse Fourier Transform of $F(\omega)$ in (2.191), we get

$$f(t) = T \sum_{k=-\infty}^{\infty} f(kT) \cdot h(t-kT) = T \sum_{k=-\infty}^{\infty} f(kT) \cdot \frac{1}{T} \operatorname{sinc} \omega_0(t-kT)$$
$$= \sum_{k=-\infty}^{\infty} f(kT) \cdot \operatorname{sinc} \omega_0(t-kT) \tag{2.193}$$

Sampling Theorem

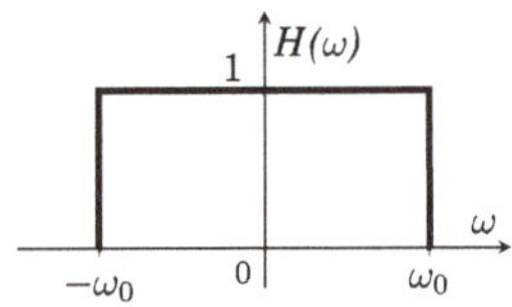

Figure 2.69: $H(\omega)$

$$h(t) = \frac{1}{2\pi}\int_{-\omega_0}^{\omega_0} 1 \cdot e^{j\omega t}\, dt = \frac{1}{2\pi}\frac{e^{j\omega t}}{+jt}\bigg|_{-\omega_0}^{\omega_0} = \frac{e^{j\omega_0 t} - e^{j\omega_0 t}}{2\pi(jt)}$$
$$= \frac{\omega_0}{\pi}\frac{\sin \omega_0 t}{t\omega_0} = \frac{\omega_0}{\pi} \operatorname{sinc} \omega_0 t = \frac{1}{T} \operatorname{sinc} \omega_0 t \tag{2.194}$$

since

$$T = \frac{\pi}{\omega_0}$$

Sampling Theorem

$$f(t) = \sum_{-\infty}^{\infty} f(kT)\,\mathrm{sinc}(\omega_0(t - kT)) \tag{2.195}$$

$f(t)$ is the continuous signal and $f(kT)$ are the signal samples at $t = kT$, where

$$T = \frac{\pi}{\omega_0} = \frac{\pi}{2\pi f_0} = \frac{1}{2f_0} \tag{2.196}$$

Notice that to avoid aliasing (overlapping) in the frequency domain in Fig. 2.67, we need

$$\omega_0 = \frac{\pi}{T} \geq B_0 \tag{2.197}$$

$$\text{Actual Sampling Interval, } T \leq \frac{\pi}{B_0} = T_N, \text{ the Nyquist Rate} \tag{2.198}$$

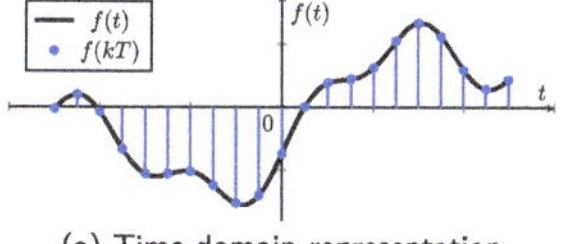

(a) Time domain representation

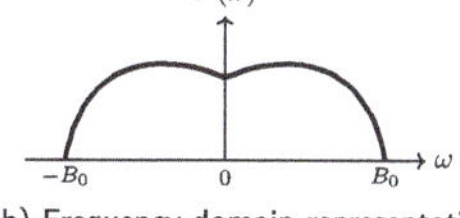

(b) Frequency domain representation

Figure 2.70: Continuous band-limited signal in time domain and in frequency domain

Sampling Theorem

In frequency domain as depicted in Fig. 2.70b, B_0 is the bandwidth of the signal, i.e.,

$$B_0 = 2\pi f_0 \tag{2.199}$$

$$T \leq \frac{1}{2f_0} = T_N \tag{2.200}$$

So long as

$$T \leq T_N = \frac{1}{2f_0} \quad \text{i.e.,} \quad \frac{1}{T} \geq 2f_0, \tag{2.201}$$

then any band-limited signal can be reconstructed from its samples using (2.195).

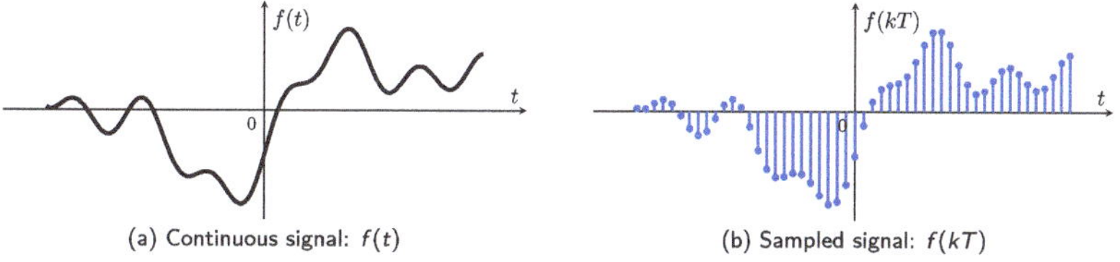

(a) Continuous signal: $f(t)$ (b) Sampled signal: $f(kT)$

Figure 2.71: Continuous and sampled signals

Sampling Theorem

To reconstruct $f(t)$, using (2.202) we need both past and future samples.

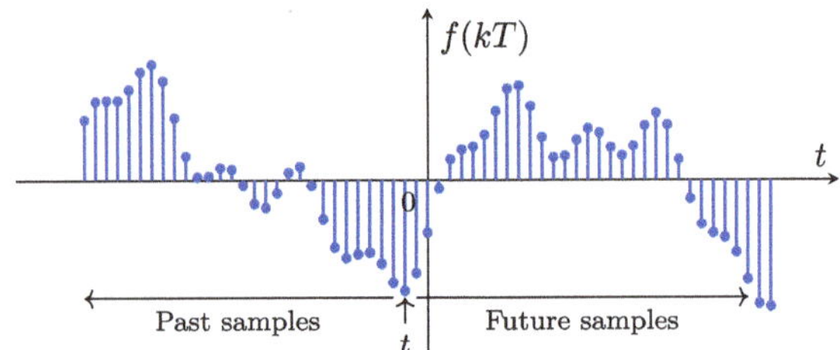

Figure 2.72: Sampled signal: $f(kT)$

$$f(t) = \sum_{k=-\infty}^{\infty} f(kT)\,\text{sinc}\,\omega_0(t-kT), \quad T \leq T_N = \frac{1}{2f_0} \tag{2.202}$$

Recovery from Past Samples

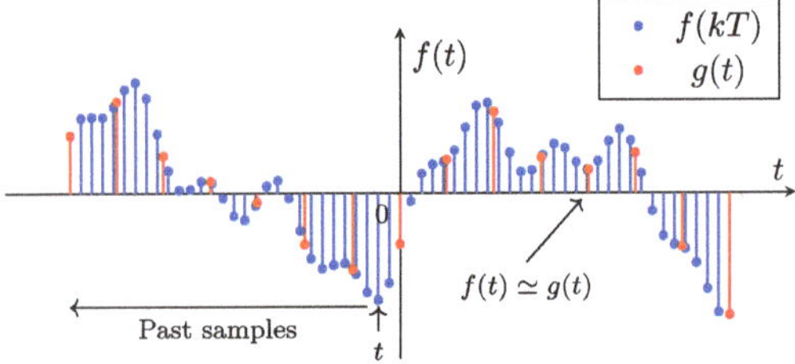

Figure 2.73: Recovered signal from past samples

Let's say we want to reconstruct $f(t)$ from n past samples. We want

$$g(t) \simeq f(t) \tag{2.203}$$

where $g(t)$ is a linear combination of past samples of $f(t)$. Thus

$$g(t) = \sum_{k=1}^{n} a_k f(t-kT) \tag{2.204}$$

where

$$a_k = (-1)^{k-1}\binom{n}{k} \tag{2.205}$$

Recovery from Past Samples

This gives

$$\mathcal{F}\{g(t)\} = G(\omega) = \sum_{k=1}^{n} a_k F(\omega) e^{-jk\omega T} \tag{2.206}$$

Error Signal

$$e(t) = f(t) - g(t) \quad \text{, where} \tag{2.207}$$

$f(t)$ is the actual signal and $g(t)$ is the signal recovered from n past samples of $f(t)$.

$$\mathcal{F}\{e(t)\} = E(\omega) = F(\omega) - G(\omega) \tag{2.208}$$

Error signal Energy:

$$\int_0^{\infty} |e(t)|^2 \, dt = \frac{1}{2\pi} \int_{-\infty}^{\infty} |E(\omega)|^2 \, d\omega \tag{2.209}$$

$$\begin{aligned} E(\omega) &= F(\omega) - \sum_{k=1}^{n} a_k e^{-jk\omega T} F(\omega) = F(\omega) \underbrace{\left[1 - \sum_{k=1}^{n} a_k e^{-jk\omega T}\right]}_{H(\omega)} \\ &= F(\omega)\, H(\omega) \end{aligned} \tag{2.210}$$

Recovery from Past Samples

Using (2.205),

$$\begin{aligned} H(\omega) &= 1 - \sum_{k=1}^{n} a_k e^{-jk\omega T} = 1 - \sum_{k=1}^{n} (-1)^{k-1} \binom{n}{k} e^{-jk\omega T} \\ &= \sum_{k=0}^{n} (-1)^k \binom{n}{k} e^{-jk\omega T} = \left(1 - e^{j\omega T}\right)^n \end{aligned} \tag{2.211}$$

$$|H(\omega)| = \left(1 - e^{j\omega T}\right)^n = \left| e^{\frac{-j\omega n T}{2}} \right| \left(e^{\frac{+j\omega T}{2}} - e^{\frac{-j\omega T}{2}} \right)^n \tag{2.212}$$

where

$$\left(e^{\frac{+j\omega T}{2}} - e^{\frac{-j\omega T}{2}} \right)^n = \left| 2j \sin \frac{\omega T}{2} \right|^n \tag{2.213}$$

Therefore

$$|H(\omega)|^2 = \left[2 \sin\left(\frac{\omega T}{2}\right) \right]^n \tag{2.214}$$

Recovery from Past Samples

Using (2.214) in (2.209) - (2.210), we get

$$\int_{-\infty}^{\infty} |e(t)|^2 = \frac{1}{2\pi} \int_{-B_0}^{B_0} |F(\omega)|^2 \left[2\sin\left(\frac{\omega T}{2}\right)\right]^n d\omega \tag{2.215}$$

Since $F(\omega)$ is arbitrary, the error signal energy on the left to go to zero, we must have

$$2\sin\left(\frac{\omega T}{2}\right) < 1 \quad \text{for all } |\omega| < B_0 \tag{2.216}$$

i.e.,

$$\sin\left(\frac{B_0 T}{2}\right) < \frac{1}{2}, \tag{2.217}$$

from the monotone nature of $\sin\omega$. Thus, we must have

$$\left[2\sin\left(\frac{\omega T}{2}\right)\right]^n \to 0$$

Fourier Series Fourier Transforms Amplitude Modulation Angle Modulation Digital Modulation Techniques

Recovery from Past Samples

For that we must have

$$\sin\left(\frac{B_0 T}{2}\right) < \frac{1}{2} = \sin\frac{\pi}{6} \Longrightarrow \frac{B_0 T}{2} < \frac{\pi}{6} \Longrightarrow T < \frac{\pi}{3B_0} = \frac{T_N}{3} \tag{2.218}$$

since the Nyquist sampling rate $T_N = \frac{\pi}{B_0}$.

In summary, a bandlimited signal can be reconstructed from its past samples alone, provided the sampling rate exceeds three times the Nyquist ratio.

Fundamentals of Communication Theory U. Pillai & A. Patel

3 Amplitude Modulation

Abstract

The concept of transforming low frequency signals such as audio to high frequency signals through modulation for transmission purposes is explored in this chapter. It begins with the principles of linear modulation, showing how carrier signals are combined with message signals to form modulated waveforms. In this context, Amplitude Modulation (AM) is explored in its various forms – Double Sideband Modulation (DSB), Single Sideband Modulation (SSB), and Vestigial Sideband Modulation (VSB). Different demodulation schemes are then introduced, highlighting practical receiver architectures and detection techniques for recovering the original information. A popular case of Vestigial Sideband Modulation, a compromise between DSB and SSB is also discussed, which is widely used in broadcast and communication systems. By covering both modulation and demodulation strategies, this chapter establishes the fundamental role of AM in the broader context of traditional communication theory.

U. Pillai and A. Patel, *Fundamentals of Communication Theory*,
https://doi.org/10.1007/978-3-032-20615-2_3

Table of Contents

1. Fourier Series

2. Fourier Transforms

3. Amplitude Modulation
3.1 Linear Modulation
3.2 AM Demodulation Schemes
3.3 Vestigial Sideband Modulation
3.4 Detection/Receiver

4. Angle Modulation

5. Digital Modulation Techniques

Linear Modulation

Linear Modulation refers to linearly shifting the frequency information content from a low frequency region to high frequency region. Amplitude Modulation (AM), Double Sideband Modulation (DSB) and Single Sideband Modulation (SSB) are variations of this theme.

If $f(t)$ has a low-frequency domain representation as in Fig. 3.1a then $x(t) = f(t)\cos\omega_0 t$ will have the representation as in Fig. 3.1b. Here $x(t)$ is the transmit signal, where the message $f(t)$ is modulated by the carrier $\cos\omega_0 t$.

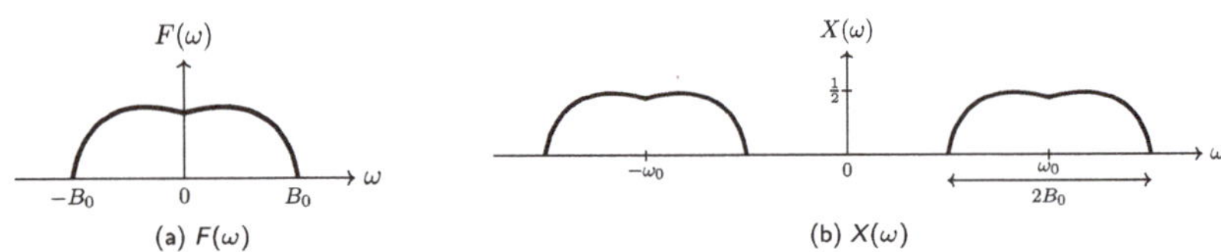

(a) $F(\omega)$ (b) $X(\omega)$

Figure 3.1: Message and Transmit signals in frequency domain

Amplitude Modulation

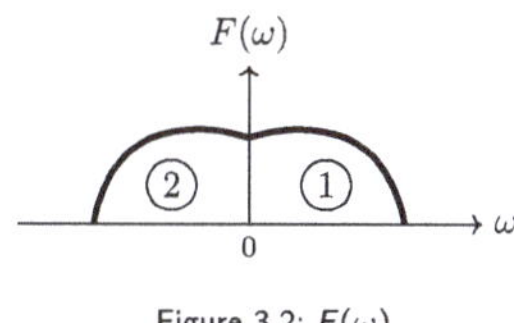

Figure 3.2: $F(\omega)$

Fig. 3.2 shows the upper ① and lower ② sidebands in the frequency domain. Presence of both of these sidebands in the frequency domain after modulation as in Fig. 3.3 is referred to as Double Sideband Modulation (DSB Modulation).

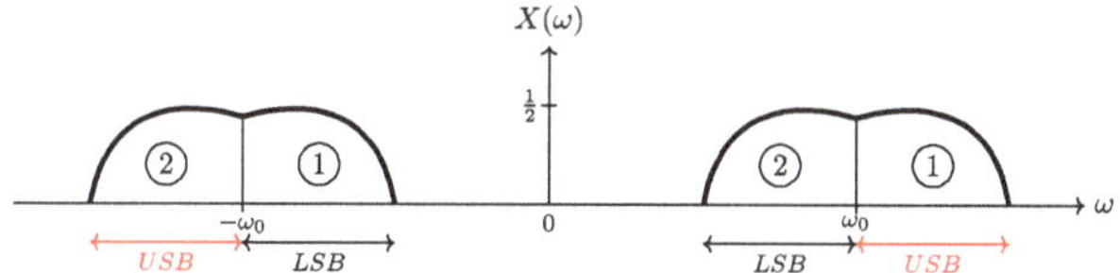

Figure 3.3: $X(\omega)$

Fourier Series Fourier Transforms Amplitude Modulation Angle Modulation Digital Modulation Techniques

Amplitude Modulation

In general, $f(t)$ can go both positive and negative and that generates ambiguity after amplitude modulation as shown in Fig. 3.4. To avoid this, a d.c. term (constant) is added to generate a non-negative message prior to modulation.

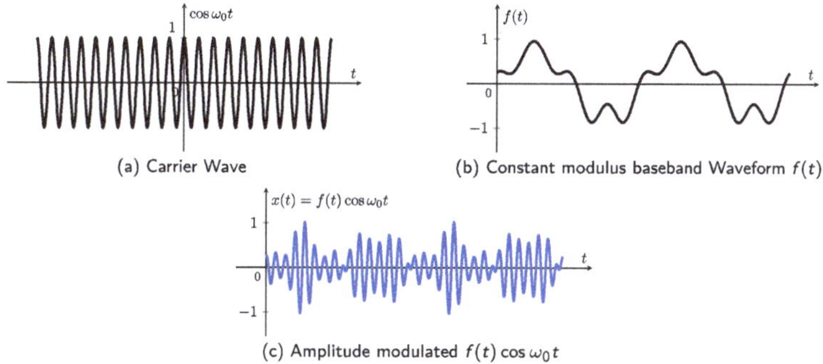

Figure 3.4: Amplitude Modulation

Fundamentals of Communication Theory U. Pillai & A. Patel

Amplitude Modulation

Presence of an additional carrier suitably scaled solves this problem

$$
\begin{aligned}
y(t) &= A\,\underbrace{[1+\overbrace{\lambda f(t)}^{\leq 1}]}_{\geq 0}\cos\omega_0 t \\
&= \underbrace{A\cos\omega_0 t}_{\text{Carrier}} + \underbrace{A\lambda f(t)\cos\omega_0 t}_{\text{DSB}}
\end{aligned}
\tag{3.1}
$$

where,

$$\lambda f(t) \leq 1, \text{ for all t} \Longrightarrow \lambda \leq \frac{1}{f(t)} = \frac{1}{m}, \text{ m} = \max\{\text{f(t)}\}$$

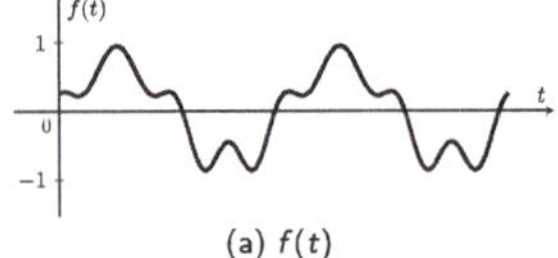

(a) $f(t)$

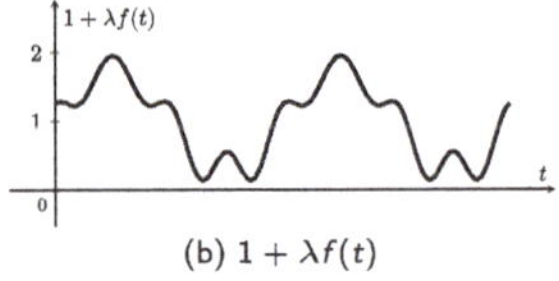

(b) $1+\lambda f(t)$

Figure 3.5: Modified message signal

Amplitude Modulation

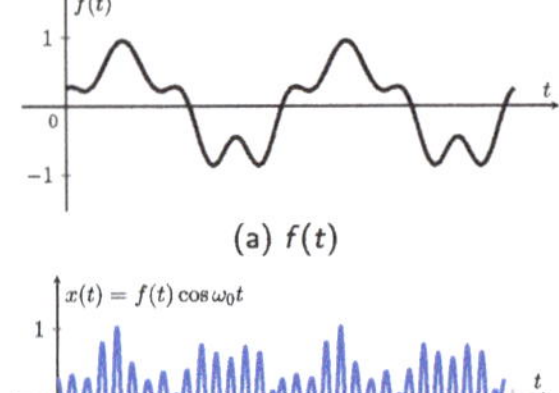

(a) $f(t)$

(c) $x(t) = f(t)\cos\omega_0 t$

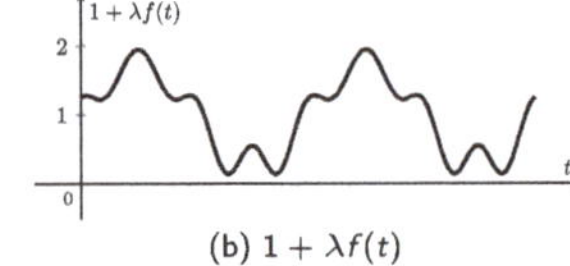

(b) $1+\lambda f(t)$

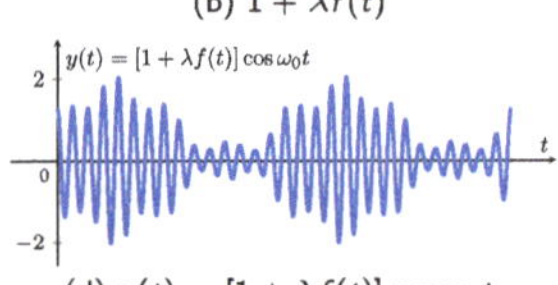

(d) $y(t) = [1+\lambda f(t)]\cos\omega_0 t$

Figure 3.6: Amplitude Modulation

$$y(t) = [1+\lambda f(t)]\cos\omega_0 t = \cos\omega_0 t + \lambda f(t)\cos\omega_0 t \tag{3.2}$$

If $f(t)$ is the message signal and has a fourier transform $F(\omega)$, then $f(t)\cos\omega_0 t$ which is the message signal modulated with the carrier $\cos\omega_0 t$ has the fourier transform

$$
\begin{aligned}
f(t) &\longleftrightarrow F(\omega) \\
f(t)\cos\omega_0 t &\longleftrightarrow \frac{1}{2}[F(\omega-\omega_0)+F(\omega+\omega_0)]
\end{aligned}
\tag{3.3}
$$

Amplitude Modulation

Double Sideband Modulation (DSB)

$$f_1(t) = A_1 \underbrace{f(t)\cos\omega_0 t}_{\text{message}}$$
$$F_1(\omega) = A_1\,\mathcal{F}\{f(t)\cos\omega_0 t\} \tag{3.4}$$

Amplitude Modulation (AM)

$$f_2(t) = A_2\big(\underbrace{\cos\omega_0 t}_{\text{carrier}} + \underbrace{m\,f(t)\cos\omega_0 t}_{\text{DSB}}\big)$$
$$F_2(\omega) = A_2\mathcal{F}\{\cos\omega_0 t\} + A_2\,m\,\mathcal{F}\{f(t)\cos\omega_0 t\} \tag{3.5}$$
$$= \frac{A_2}{2}\Big[F(\omega+\omega_0) + F(\omega-\omega_0)\Big]$$

since

$$\cos\omega_0 t = \frac{1}{2}\left(e^{j\omega_0 t} + e^{-\omega_0 t}\right) \leftrightarrow \pi[\delta(\omega+\omega_0) + \delta(\omega-\omega_0)]$$

Amplitude Modulation

Double Sideband Modulation (DSB)

$$f_1(t) = A_1 f(t)\cos\omega_0 t \tag{3.6}$$

Amplitude Modulation (AM): AM = Carrier + DSB

$$f_2(t) = A_2[1 + \lambda f(t)]\cos\omega_0 t \tag{3.7}$$

$$m = max|f(t)| \quad , \quad \lambda \leq \frac{1}{m}$$

$$P_{DSB} = \frac{1}{T_0}\int_0^{T_0} f_1^2(t)\,dt = \frac{A_1^2 m^2}{T_0}\int_{t_0}^{t_0+T_0}\left(\frac{1+\cos 2\omega_0 t}{2}\right)dt$$
$$= \frac{A_1^2 m^2}{2} + \underbrace{\frac{\sin 2\omega_0 t}{2\omega_0}\Bigg|_0^{T_0=2\pi m}}_{0} = \frac{{A_1}^2 m^2}{2} \tag{3.8}$$

Amplitude Modulation

$$\begin{aligned} f_2(t) &= A_2[1 + \lambda \underbrace{f(t)}_{\leq m}] \cos \omega_0 t \\ &= A_2(1 + \underbrace{\lambda m}_{\leq 1}) \cos \omega_0 t \leq 2A_2 \cos \omega_0 t \end{aligned} \tag{3.9}$$

Average Power

$$P_{AM} \leq \frac{1}{T_0} \int_0^{B_0} f_2^2(t)\, dt = \frac{4A_2^2}{T_0} \int_0^{T} \left(\frac{1 + \cos 2\omega_0 t}{2} \right) dt = 2A_2^2 \tag{3.10}$$

Thus for power parity between DSB and AM, we must have

$$P_{AM} - P_{DSB} \longrightarrow 2A_2^2 = \frac{A_1^2 m^2}{2} \tag{3.11}$$

$$A_1\, m = 2A_2 \Longrightarrow DSB \simeq AM \tag{3.12}$$

Amplitude Modulation

DSB has an improvement over AM in power since DSB has no carrier.
For real $f(t)$,

$$\begin{aligned} F(\omega) &= \int_{-\infty}^{\infty} f(t) e^{-j\omega t}\, dt \\ &= \underbrace{\int_{-\infty}^{\infty} f(t) \cos \omega_0 t\, dt}_{\text{Real part, } R(\omega)} - \underbrace{j \int_{-\infty}^{\infty} f(t) \sin \omega_0 t\, dt}_{\text{Imaginary part, } jX(\omega)} \end{aligned} \tag{3.13}$$

$$\begin{aligned} \text{Even, } R(\omega) &= \int_{-\infty}^{\infty} f(t) \cos \omega_0 t\, dt = R(-\omega) \\ \text{Odd, } X(\omega) &= \int_{-\infty}^{\infty} f(t) \sin \omega_0 t\, dt = -X(-\omega) \end{aligned} \tag{3.14}$$

Single Sideband Modulation

Consider

$$\underbrace{f_3(t)}_{\text{real}} = \underbrace{f(t)}_{\text{real}} \cos\omega_0 t + \underbrace{\hat{f}(t)}_{\text{real}} \sin\omega_0 t \tag{3.15}$$

Hilbert Transform,

$$\underbrace{\hat{f}(t)}_{\text{real}} = f(t) \circledast \frac{1}{\pi t} = \frac{1}{\pi}\int_{-\infty}^{\infty} \frac{f(\tau)}{t-\tau}\, d\tau \tag{3.16}$$

$$\hat{F}(\omega) = \underbrace{-j\,\mathrm{sgn}\,\omega}_{H(\omega)}\, F(\omega) \tag{3.17}$$

where,

$$f(t) \longleftrightarrow F(\omega)$$

$$f(t)\cos\omega_0 t \longleftrightarrow \frac{1}{2}[F(\omega+\omega_0) + F(\omega-\omega_0)]$$

$$\hat{f}(t)\sin\omega_0 t \longleftrightarrow \frac{1}{2}[F(\omega+\omega_0)\,\mathrm{sgn}(\omega+\omega_0) - F(\omega-\omega_0)\,\mathrm{sgn}(\omega-\omega_0)]$$

so that (3.18)

$$F_3(\omega) = \frac{1}{2}[(1+\mathrm{sgn}(\omega+\omega_0)F(\omega+\omega_0)) + (1-\mathrm{sgn}(\omega-\omega_0)F(\omega-\omega_0))] \tag{3.19}$$

Fourier Series Fourier Transforms Amplitude Modulation Angle Modulation Digital Modulation Techniques

Single Sideband Modulation

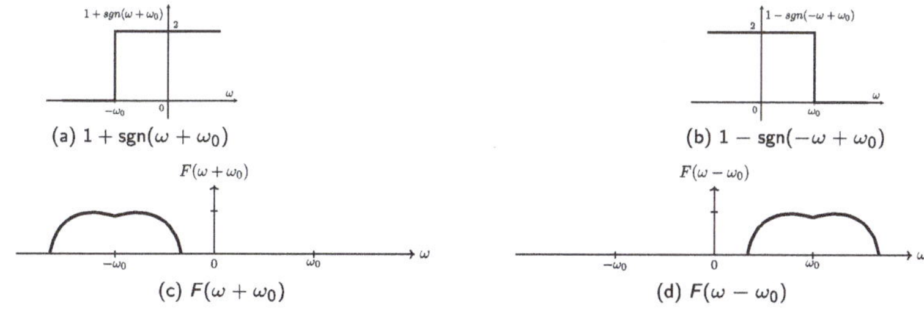

(a) $1 + \mathrm{sgn}(\omega + \omega_0)$ (b) $1 - \mathrm{sgn}(-\omega + \omega_0)$

(c) $F(\omega + \omega_0)$ (d) $F(\omega - \omega_0)$

Figure 3.7: Generating SSB modulated signal

Multiplying signals in (a) and (b) with (c) and (d) respectively and adding them results in:

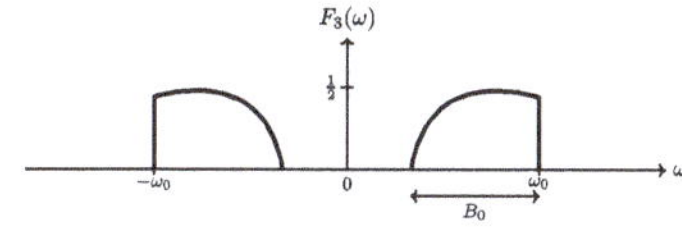

Figure 3.8: SSB modulation(LSB) in frequency domain

Single Sideband Modulation

$$f(t)\cos\omega_0 t \longleftrightarrow \frac{1}{2}[F(\omega-\omega_0)+F(\omega+\omega_0)]$$
$$\hat{f}(t)\sin\omega_0 t \longleftrightarrow \frac{1}{2j}[\hat{F}(\omega-\omega_0)-\hat{F}(\omega+\omega_0)] \tag{3.20}$$

$$\hat{F}(\omega) = -j\ \text{sgn}\,\omega F(\omega)$$
$$f_4(t) = f(t)\cos\omega_0 t - \hat{f}(t)\sin\omega_0 t \tag{3.21}$$

$$f_4(t) \longleftrightarrow \frac{1}{2}[(1-\text{sgn}(\omega+\omega_0))F(\omega+\omega_0)+(1+\text{sgn}(\omega-\omega_0))F(\omega-\omega_0)] \tag{3.22}$$

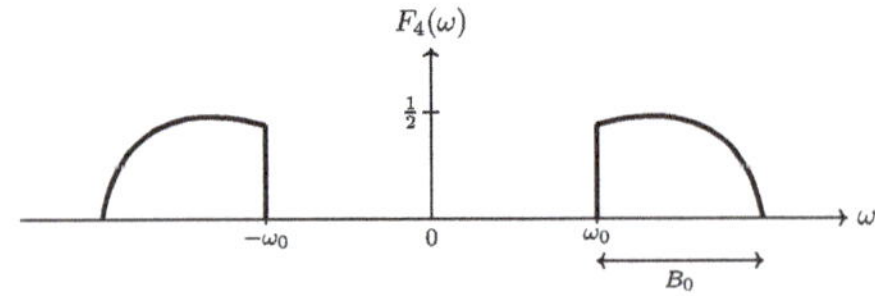

Figure 3.9: $F_4(\omega)$: SSB modulation(USB) in frequency domain

Fourier Series Fourier Transforms Amplitude Modulation Angle Modulation Digital Modulation Techniques

Single Sideband Modulation

Average Power for SSB modulation

$$f_3(t) = \frac{A_1}{2}[f(t)\cos\omega_0 t \pm \hat{f}(t)\sin\omega_0 t] \tag{3.23}$$

$$P_{SSB} = 2\cdot\left(\frac{A_1}{2}\right)^2 \cdot \underbrace{\frac{1}{T_0}\int_0^{T_0} f^2(t)\cos^2\omega_0 t\,dt}_{P_0\text{: Average Power}} = \frac{A_1^2}{2}P_0 \tag{3.24}$$

$$P_{DSB} = A_1^2 \cdot \underbrace{\frac{1}{T_0}\int_0^{T_0} f^2(t)\cos^2\omega_0 t\,dt}_{P_0} = A_1^2 P_0 \tag{3.25}$$

Fundamentals of Communication Theory U. Pillai & A. Patel

Single Sideband Modulation

To save bandwidth, SSB modulation is preferred over DSB modulation.

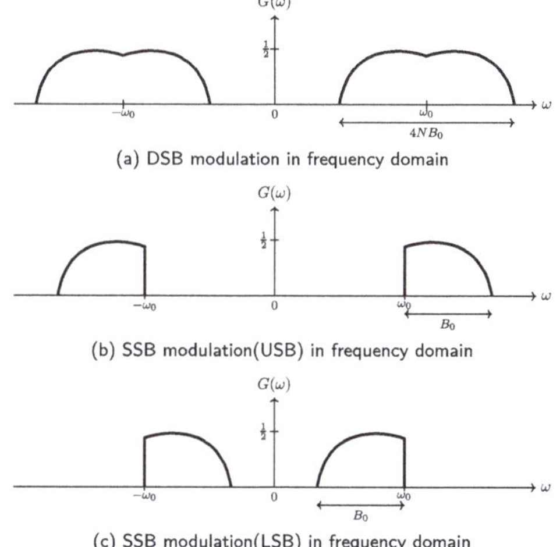

(a) DSB modulation in frequency domain

(b) SSB modulation(USB) in frequency domain

(c) SSB modulation(LSB) in frequency domain

Figure 3.10: Frequency domain representation of Double and Single Sideband Modulation

AM Demodulation Schemes - Envelope Detector

Information bearing signal $f(t)$ is in the envelope of the AM signal. Use a low-pass filter to extract the envelope as shown in Fig. 3.11

$$\text{AM: } f_2(t) = [1 + \underbrace{\lambda f(t)}_{\leq 1}] \cos \omega_0 t = y(t) \tag{3.26}$$

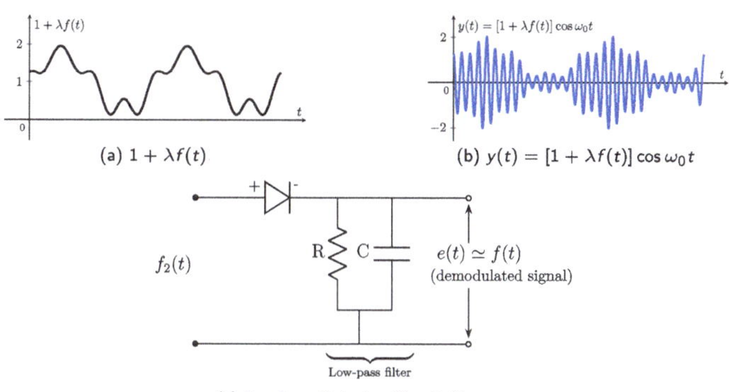

(a) $1 + \lambda f(t)$

(b) $y(t) = [1 + \lambda f(t)] \cos \omega_0 t$

(c) Envelope Detector Circuit Diagram

Figure 3.11: Envelope Detector to demodulate an AM signal

AM Demodulation Schemes - Envelope Detector

In Fig. 3.11, the filter output $e(t)$ represents the demodulated signal. Diode conducts when $f_2(t)$ exceeds the output $e(t)$, otherwise the output $e(t)$ follows $f_2(t)$. Since the diode conducts upto $t = t_0$,

$$e(t_0) = 1 + \lambda f(t_0) \tag{3.27}$$

which is same as the envelope voltage.
For $t > t_0$, since $f_2(t) < e(t)$, the diode does not conduct and the capacitor discharges as

$$e(t) = e(t_0)\, e^{-(t-t_0)/RC} = [1 + \lambda f(t_0)]\, e^{-(t-t_0)/RC} \tag{3.28}$$

This continues till $t = t_1$, after which since $f_2(t) > e(t)$, the diode starts conducting again and the output $e(t)$ rises with $f_2(t)$ till $t = t_2$ etc. Thus the output $e(t)$ follows the envelope $1 + \lambda f(t_0)$, provided the capacitor does not discharge too slowly or too quickly.

AM Demodulation Schemes - Envelope Detector

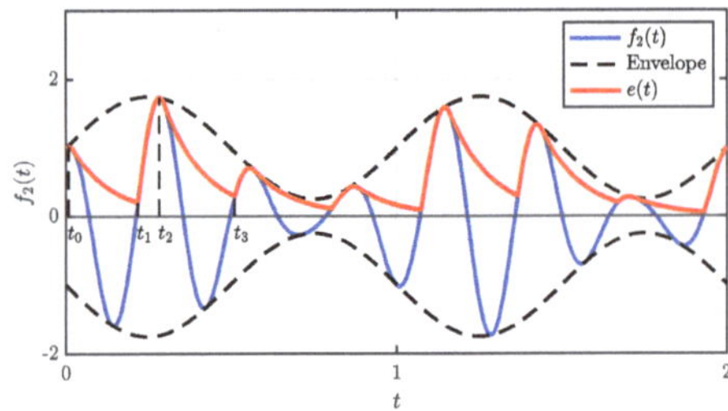

Figure 3.12: The clipping effect

$$f_2(t) = [1 + \lambda f(t)] \cos \omega_0 t \tag{3.29}$$

Demodulated signal,

$$e(t) \simeq \lambda f(t) \tag{3.30}$$

since

$$\begin{aligned} e(t) &= f_2(t_0) e^{-(t-t_0)/RC}, \quad t_0 < t < t_1. \\ f_2(t_0) &= [1 + \lambda f(t_0)] \end{aligned} \tag{3.31}$$

AM Demodulation Schemes - Envelope Detector

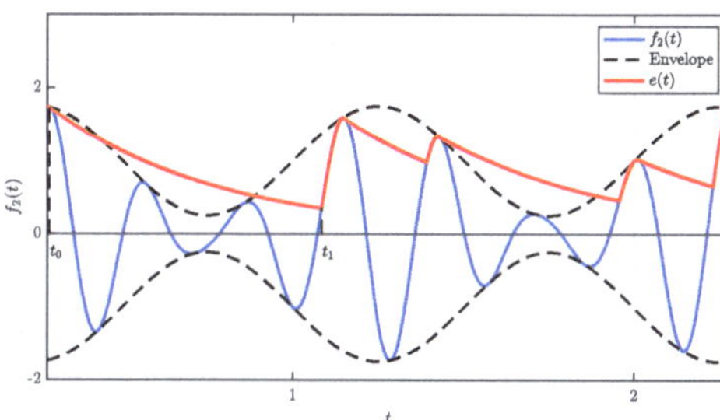

Figure 3.13: The envelope detector

Ideally, the output $e(t)$ should hug the envelope $[1 + \lambda f(t)]$. Adjust values of RC accordingly to avoid clipping effect caused by slow capacitor delay as shown in Fig. 3.12

If C is not properly chosen, improper clipping will result in lost message information since the capacitor would discharge too slowly.

AM Demodulation Schemes - Envelope Detector

To avoid clipping, rate of change of the message should be approximately equal to the rate of change of capacitor output $e(t)$, i.e.,

$$\frac{d}{dt}(1 + \lambda f(t))\Big|_{t=t_0} = \frac{d}{dt}e(t)\Big|_{t=t_0} \tag{3.32}$$

$$e(t) = [1 + \lambda f(t_0)]e^{-(t-t_0)/RC} \tag{3.33}$$

$$\frac{de(t)}{dt}\Big|_{t=t_0} = \frac{[1 + \lambda f(t_0)]}{-RC}e^{-(t-t_0)/RC}\Big|_{t=t_0} \\ \Rightarrow -\frac{1 + \lambda f(t_0)}{RC} = \lambda\frac{df(t_0)}{dt} \tag{3.34}$$

Or,

$$-\frac{1}{RC} = \frac{\lambda}{1 + \lambda f(t_0)} \cdot \frac{df(t_0)}{dt} \tag{3.35}$$

Therefore, to avoid clipping

$$-\frac{1}{RC} \simeq \frac{\lambda}{1 + \lambda f(t_0)} \cdot \frac{df(t_0)}{dt} \tag{3.36}$$

AM Demodulation Schemes - Envelope Detector

Example

$$\begin{aligned} f(t) &= \cos B_0 t \\ f'(t) &= -B_0 \sin B_0 t \end{aligned} \tag{3.37}$$

$$\frac{1}{RC} \simeq \frac{\lambda}{1+\lambda \cos B_0 t_0} B_0 \sin B_0 t \tag{3.38}$$

$$\frac{1}{B_0 \lambda RC} = \frac{\sin B_0 t_0}{1+\lambda \cos B_0 t_0} \tag{3.39}$$

Let

$$\frac{\sin x}{1+\lambda \cos x} = g(x) \tag{3.40}$$

$g(x)$ is maximum when $g'(x) = 0$.

$$g(x) = \frac{\sin x}{1+\lambda \cos x} \tag{3.41}$$

Fourier Series Fourier Transforms Amplitude Modulation Angle Modulation Digital Modulation Techniques

AM Demodulation Schemes - Envelope Detector

$$g'(x) = \frac{(1+\lambda \cos x)(\cos x) - \sin x(-\lambda \sin x)}{(1+\lambda \cos x)^2} \tag{3.42}$$

$$\begin{aligned} g'(x) &= \frac{\lambda + \cos x}{(1+\lambda \cos x)^2} = 0 \Rightarrow \cos x_0 = -\lambda \\ &\Rightarrow \sin x_0 = \sqrt{1-\lambda^2} \end{aligned} \tag{3.43}$$

$$g(\cos x = -\lambda) = \frac{\sqrt{1-\lambda^2}}{1-\lambda^2} = \frac{1}{\sqrt{1-\lambda^2}} \tag{3.44}$$

From (3.39)

$$\begin{aligned} \frac{1}{B_0 \lambda RC} &\simeq \frac{1}{\sqrt{1-\lambda^2}} \\ RC &\simeq \frac{\sqrt{1-\lambda^2}}{B_0 \lambda} \end{aligned} \tag{3.45}$$

Fundamentals of Communication Theory U. Pillai & A. Patel

AM Demodulation Schemes

DSB Receiver

$$f_2(t) = f(t)\cos\omega_0 t \tag{3.46}$$

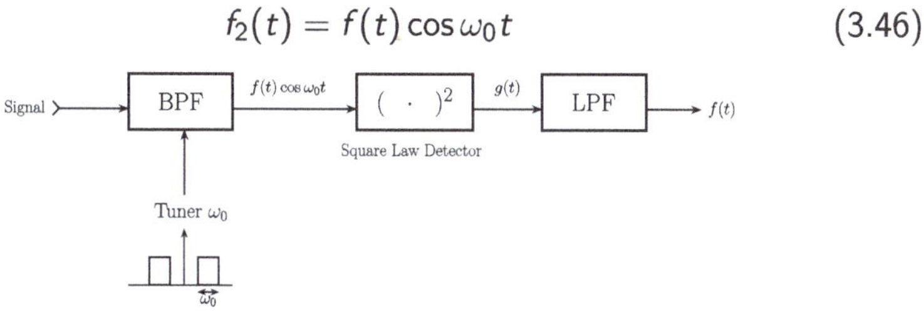

Figure 3.14: Square Law Detector

$$g(t) = [f(t)\cos\omega_0 t]^2 = f^2(t)\cos^2\omega_0 t \tag{3.47}$$

$$f_1(t) = [1 + \lambda f(t)]\cos\omega_0 t \tag{3.48}$$

$$\begin{aligned} g(t) &= [1 + \lambda f(t)]^2 \cos^2\omega_0 t \\ &= \left[1 + 2\lambda f(t) + \lambda^2 f^2(t)\right]\left(\frac{1 + \cos 2\omega_0 t}{2}\right) \end{aligned} \tag{3.49}$$

Fourier Series Fourier Transforms **Amplitude Modulation** Angle Modulation Digital Modulation Techniques

AM Demodulation Schemes

We know,

$$\lambda \simeq \frac{1}{M} \Longrightarrow \lambda^2 \simeq 0$$

$$g(t) = \frac{1}{2}[1 + 2\lambda f(t)](1 + \cos 2\omega_0 t) \tag{3.50}$$

From Fig. 3.15, $g(t)$ can be written as:

$$g(t) = \frac{1}{2}[1 + 2\lambda f(t)](1 + \cos 2\omega_0 t) \tag{3.51}$$

$f(t)$ can be recovered by passing $g(t)$ through a Low-Pass Filter.

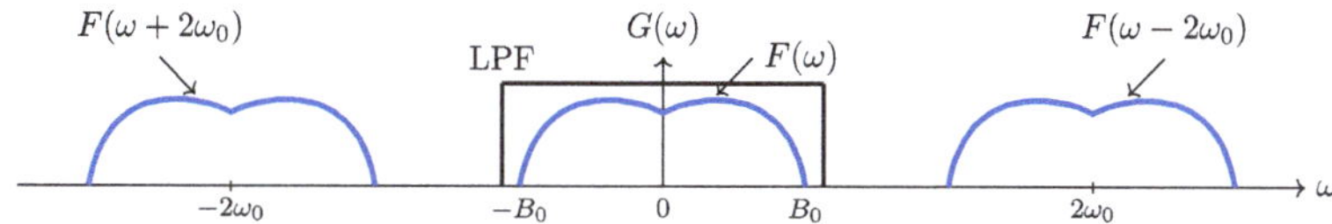

Figure 3.15: Obtaining $f(t)$ after passing through LPF

Fundamentals of Communication Theory U. Pillai & A. Patel

AM Demodulation Schemes

Synchronous Detector

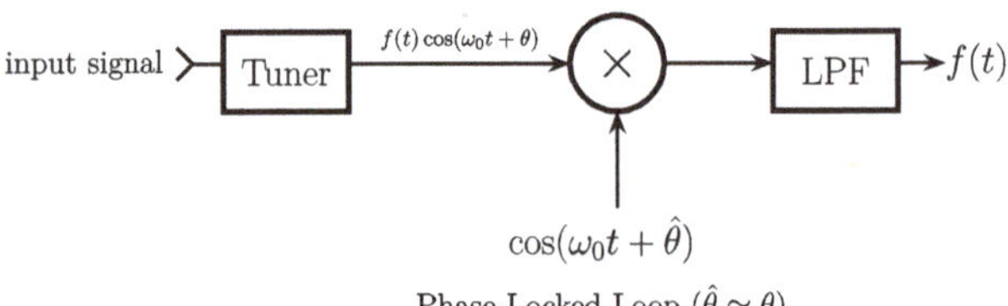

Figure 3.16: Synchronous DSB Receiver

In absence of a carrier, such as in DSB, to recover $f(t)$ from

$$f_2(t) = A_2 f(t)\cos\omega_0 t\,, \tag{3.52}$$

one needs to generate carrier waveform at the receiver that is in phase with the $\cos\omega_0 t$ waveform in the transmitted signal $f_2(t)$.

Fourier Series Fourier Transforms Amplitude Modulation Angle Modulation Digital Modulation Techniques

AM Demodulation Schemes

Asynchronous Detector

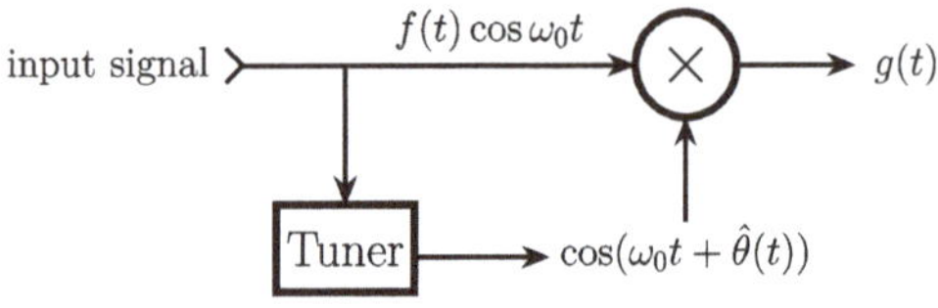

Figure 3.17: Asynchronous DSB Receiver

In Fig. 3.17,

$$\begin{aligned} g(t) &= f(t)\cos\omega_0 t.\cos(\omega_0 t+\theta) = \frac{1}{2}f(t)[\cos\theta + \cos(2\omega_0 t+\theta)] \\ &= \underbrace{\frac{1}{2}f(t)\cos\theta}_{\text{Baseband}} + \frac{1}{2}f(t)\cos\theta.\cos(2\omega_0 t+\theta) \end{aligned} \tag{3.53}$$

Passing $g(t)$ through a Low-Pass Filter, we get the output signal as $f(t)\cos\theta(t)$, where $\theta(t)$ introduces fading. To remove fading, estimate $\theta(t)$ and insert this into the receiver to match the incoming phase of the signal.

Fundamentals of Communication Theory U. Pillai & A. Patel

Vestigial Sideband Modulation

$f(t)\cos\omega_0 t + f(t)\sin\omega_0 t \circledast h(t)$

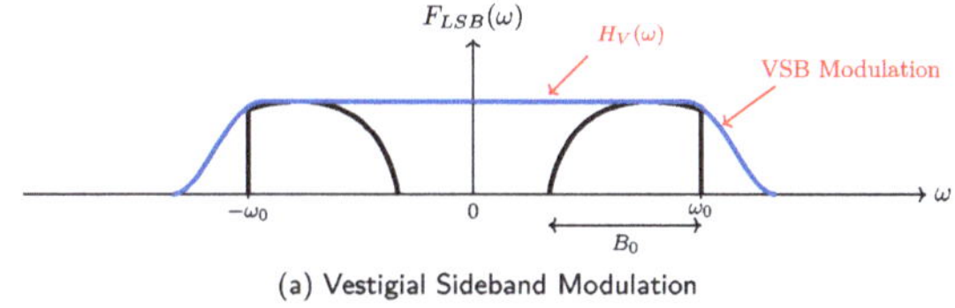

(a) Vestigial Sideband Modulation

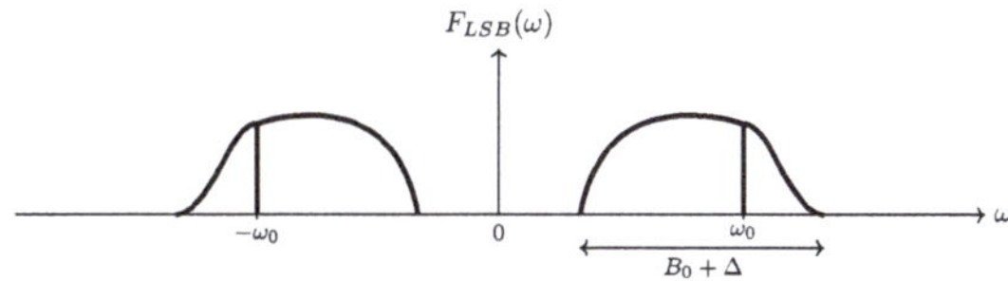

(b) $F_{USB}(\omega)$: SSB modulation(USB) in frequency domain in reality

Figure 3.18: Vestigial Sideband Modulation

Fourier Series Fourier Transforms Amplitude Modulation Angle Modulation Digital Modulation Techniques

Vestigial Sideband Modulation

Ideal SSB,

$$\underbrace{f_3(t)}_{\text{real}} = x(t) = \underbrace{f(t)}_{\text{real}}\cos\omega_0 t + \hat{f}(t)\sin\omega_0 t \tag{3.54}$$

where $\hat{f}(t)$ is the H.T. of $f(t)$ such that

$$\hat{f}(t) = \int \frac{f(\tau)}{t-\tau}\, d\tau \tag{3.55}$$

Ideal SSB is impractical

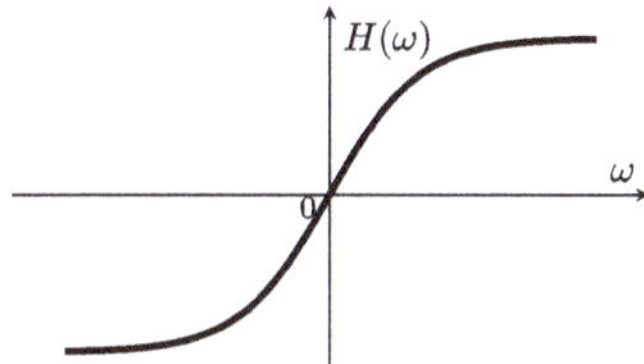

Figure 3.19: Practical Hilbert Transform

Vestigial Sideband Modulation

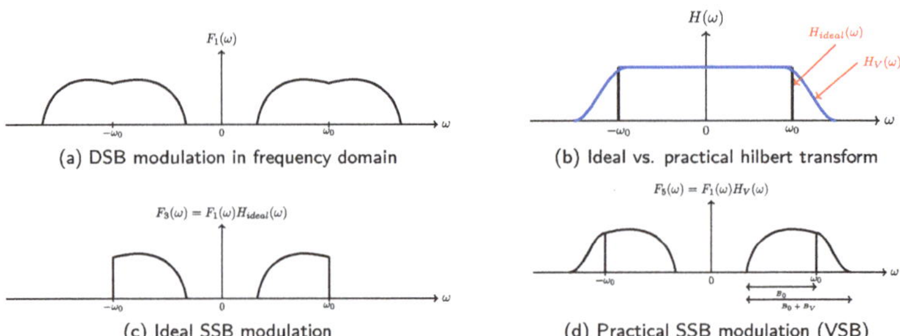

(a) DSB modulation in frequency domain

(b) Ideal vs. practical hilbert transform

(c) Ideal SSB modulation

(d) Practical SSB modulation (VSB)

Figure 3.20: SSB vs. VSB

$$F_3(\omega) = F_1(\omega)H_{ideal}(\omega) \tag{3.56}$$

$$F_5(\omega) = F_1(\omega)H_V(\omega) = [F(\omega+\omega_0) + F(\omega-\omega_0)]H_V(\omega) \tag{3.57}$$

Fourier Series Fourier Transforms Amplitude Modulation Angle Modulation Digital Modulation Techniques

Vestigial Sideband Modulation

$$\begin{aligned} f_5(t) &= \frac{1}{2\pi}\int_{-\infty}^{\infty} F_5(\omega)H_V(\omega)e^{j\omega t}\,d\omega \\ &= \int_{-\infty}^{\infty} [F(\omega+\omega_0) + F(\omega-\omega_0)]H_V(\omega)e^{j\omega t}\,d\omega \end{aligned} \tag{3.58}$$

Substituting $\omega + \omega_0$ with x and $\omega - \omega_0$ with y, we get

$$\begin{aligned} f_5(t) = &\int_{-\infty}^{\infty} F(x)H_V(x-\omega_0)e^{+j(x-\omega_0)t}\,dx + \\ &\int_{-\infty}^{\infty} F(y)H_V(y+\omega_0)e^{+j(y+\omega_0)t}\,dy \end{aligned} \tag{3.59}$$

Replacing x and y with ω, we get

$$\begin{aligned} f_5(t) = &\, e^{-j\omega_0 t}\int_{-\infty}^{\infty} F(\omega)H_V(\omega-\omega_0)e^{j\omega t}\,d\omega \\ &+ e^{j\omega_0 t}\int_{-\infty}^{\infty} F(\omega)H_V(\omega+\omega_0)e^{j\omega t}\,d\omega \end{aligned} \tag{3.60}$$

Vestigial Sideband Modulation

$$f_5(t) = \cos\omega_0 t \left[\frac{1}{2T}\int_{-B_0}^{B_0} F(\omega)(H_V(\omega+\omega_0) + H_V(\omega-\omega_0))e^{j\omega t}\, d\omega\right]$$
$$+ j\sin\omega_0 t \left[\frac{1}{2T}\int_{-B_0}^{B_0} F(\omega)(H_V(\omega+\omega_0) + H_V(\omega-\omega_0))e^{j\omega t}\, d\omega\right] \tag{3.61}$$

$$H_V(\omega+\omega_0) + H_V(\omega-\omega_0) \equiv 1, |\omega| < B_0 \tag{3.62}$$

$$f_5(t) = f(t)\cos\omega_0 t + \frac{j\sin\omega_0 t}{2\pi}\int_{-B_0}^{B_0} F(\omega)[H_V(\omega+\omega_0) - H_V(\omega-\omega_0)]e^{j\omega t} \tag{3.63}$$

$$\underbrace{K_q(t)}_{\text{real}} \leftrightarrow K(\omega) = j\,\underbrace{H_V(\omega+\omega_0) - H_V(\omega-\omega_0)}_{\text{odd function}} \tag{3.64}$$

Vestigial Sideband Modulation

$$H_V(\omega) = R(\omega) + jX(\omega) = R(-\omega) - jX(-\omega) \tag{3.65}$$

$$H_V(\omega+\omega_0) + H_V(\omega-\omega_0) = 1 \tag{3.66}$$

$$R(\omega+\omega_0) + jX(\omega+\omega_0) + R(\omega-\omega_0) + jX(\omega-\omega_0) = 1 \tag{3.67}$$

$$R(\omega_0+\omega) + R(\omega_0-\omega) = 1 \tag{3.68}$$

$$R(\omega+\omega_0) + R(\omega-\omega_0) = 1 \tag{3.69}$$

$$X(\omega+\omega_0) + X(\omega-\omega_0) = 0 \tag{3.70}$$

$$X(\omega_0+\omega) = -X(\omega-\omega_0) = X(\omega_0-\omega) \tag{3.71}$$

Vestigial Sideband Modulation

$$f_5(t) = f(t)\cos\omega_0 t + \frac{j}{2\pi}\int_{-B_0}^{B_0} F(\omega)[H_V(\omega+\omega_0) - H_V(\omega-\omega_0)]e^{j\omega t}\,d\omega \tag{3.72}$$

$$\begin{aligned} K(\omega) &= H_V(\omega+\omega_0) - H_V(\omega-\omega_0) \\ &= R(\omega+\omega_0) + jX(\omega+\omega_0) - R(\omega_0-\omega) - jX(\omega-\omega_0) \\ &= R(\omega+\omega_0) - R(\omega_0-\omega) + j2X(\omega_0-\omega) \end{aligned} \tag{3.73}$$

$$K_q(t) \leftrightarrow K_q(\omega) = -K_q(-\omega) \tag{3.74}$$

Therefore, the second term becomes

$$\frac{j}{2\pi}\int_{-B_0}^{B_0} F(\omega)K_q(\omega)e^{j\omega t}\,d\omega - \underbrace{\int_{-B_0}^{B_0} F(\omega)X(\omega_0-\omega)e^{j\omega t}\,d\omega}_{=0} \tag{3.75}$$

$$= \frac{j}{2\pi}\int_{-B_0}^{B_0} \underbrace{A(\omega)}_{\text{even}}\,\underbrace{K_q(\omega)}_{\text{odd}}\,e^{j[\psi(\omega)+\omega t]}\,d\omega \tag{3.76}$$

Fourier Series Fourier Transforms Amplitude Modulation Angle Modulation Digital Modulation Techniques

Vestigial Sideband Modulation

Therefore, the second term can be rewritten as

$$\frac{j}{2\pi}\int_{-B_0}^{B_0} A(\omega)K_q(\omega)j\sin(\psi(\omega)+\omega t)\,d\omega \tag{3.77}$$

$$\begin{aligned} f_5(t) &= f(t)\cos\omega_0 t - \frac{\sin\omega_0 t}{2\pi}\underbrace{\int A(\omega)K_q(\omega)\sin(\psi(\omega)+\omega t)\,d\omega}_{\longleftrightarrow f_q(t) = f(t)\circledast K_q(t)} \\ &= f(t)\cos\omega_0 t - \underbrace{f_q(t)}_{\hat{f}(t)}\sin\omega_0 t \end{aligned} \tag{3.78}$$

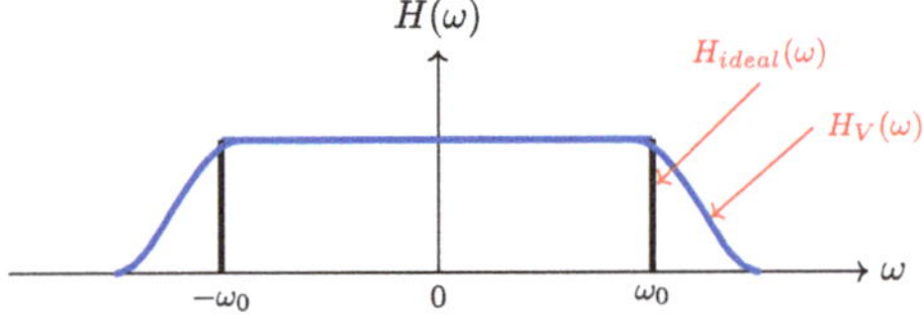

Figure 3.21: Ideal vs. practical Hilbert Transform

Vestigial Sideband Modulation

$$f_3(t) = f(t)\cos\omega_0 t - \hat{f}(t)\sin\omega_0 t \quad (3.79)$$

$$f_5(t) = f(t)\cos\omega_0 t - f_q(t)\sin\omega_0 t \quad (3.80)$$

where

$$f_q(t) = f(t) \circledast K_q(t)$$

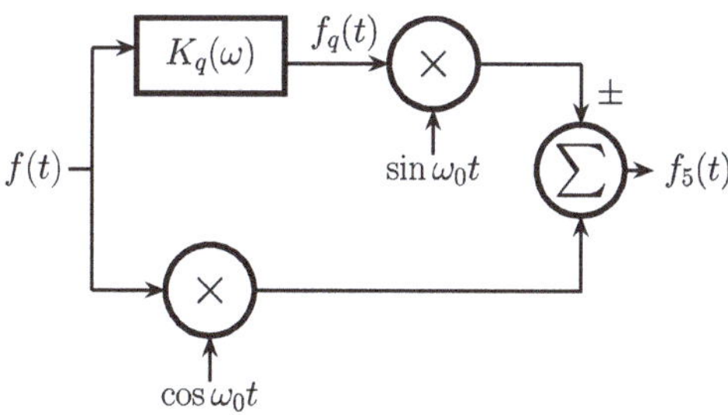

Figure 3.22: Block Diagram for VSB signal generation

Fourier Series Fourier Transforms Amplitude Modulation Angle Modulation Digital Modulation Techniques

Superheterodyne Receiver

In a superheterodyne receiver, irrespective of the incoming frequency range, the processing is carried out at some intermediate frequency ω_i.

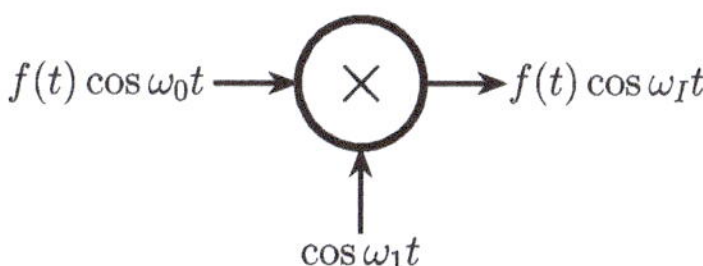

Figure 3.23: Block diagram for superheterodyne receiver

In Fig. 3.23,

$$\omega_1 = \omega_0 + \omega_I \quad \text{(Higher)}$$

or

$$\omega_1 = \omega_0 - \omega_I \quad \text{(Lower)}$$

Superheterodyne Receivers have a high dynamic range: 100 kHz to 500 kHz

Fundamentals of Communication Theory U. Pillai & A. Patel

Superheterodyne Receiver

$$\omega_1 = \omega_I + \omega_0 \tag{3.81}$$

$$f(t)\cos\omega_0 t\cos\omega_1 t = \frac{1}{2}f(t)\left[\cos\omega_I t + \cos(2\omega_0 + \omega_I)t\right] \tag{3.82}$$

After passing through a Band pass filter, the frequencies centered around ω_I would be preserved and the remaining frequencies are removed.

$$\underbrace{f(t)\cos\omega_0 t}_{\text{incoming signal}}\ \underbrace{\cos\omega_1 t}_{\text{tuner}} = \frac{1}{2}f(t)\left[\cos\underbrace{(\omega_0 - \omega_1)}_{\omega_I=\omega_0-\omega_1}t + \cos\underbrace{(\omega_0 + \omega_1)}_{\omega_1=\omega_I-\omega_0}t\right] \tag{3.83}$$

Two options for tuner frequency ω_1 :

$$\begin{aligned}\omega_1 &= \underbrace{\omega_0 - \omega_I}_{\text{lower}} < \omega_0\\ \omega_1 &= \underbrace{\omega_0 + \omega_I}_{\text{super}} > \omega_0\end{aligned} \tag{3.84}$$

Select the one with higher dynamic range.

Fourier Series Fourier Transforms Amplitude Modulation Angle Modulation Digital Modulation Techniques

Superheterodyne Receiver

$$\text{Dynamic Range, } \omega_0 = \begin{cases}110kHz\\ 200kHz\end{cases} \tag{3.85}$$

$$\omega_I = 100kHz$$

$$\begin{aligned}\omega_1 &= \omega_0 - \omega_I \to 10kHz \to 100kHz\\ &\quad or\\ \omega_1 &= \underbrace{\omega_0 + \omega_I}_{\text{super}} \to 210Hz \to 310Hz\end{aligned} \tag{3.86}$$

The superior is the larger frequency that provides a better dynamic range for ω_i.

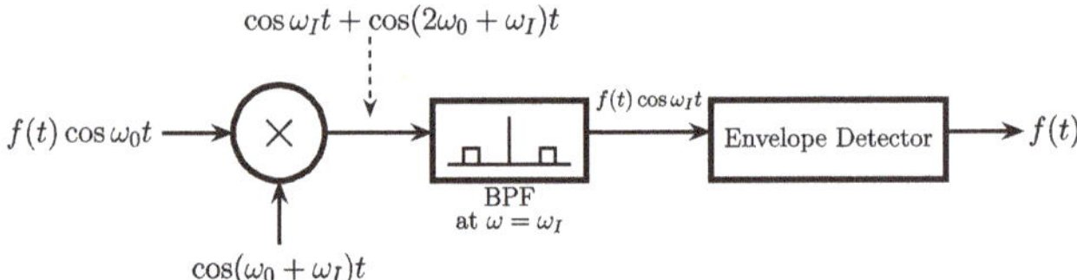

Figure 3.24: Block diagram for superheterodyne receiver

Superheterodyne Receiver

$$f_1(t)\cos(\omega_0 + 2\omega_I)t\cos(\omega_0 + \omega_I)t = f_1(t)\cos\omega_I t + f_1(t)\cos(2\omega_0 + 3\omega_I)t \quad (3.87)$$

In Fig. 3.24, if there is an unwanted signal $f_1(t)\cos(\omega_0 + 2\omega_I)t$ i.e., the mirror image frequency of ω_0, along with $f(t)\cos\omega_0 t$, then after passing it through the super-heterodyne receiver, there will be a resultant output signal,

$$[f(t) + f_1(t)]\cos\omega_I t \quad (3.88)$$

which will be passed to the envelope detector for demodulation. This should be taken into modulation constraints.

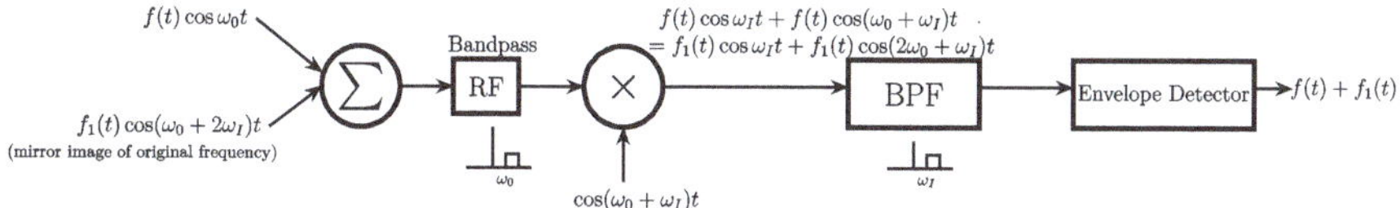

Figure 3.25: Block diagram for superheterodyne receiver with multiple input signals

Superheterodyne Receiver

In Fig. 3.25, $f_1(t)\cos(\omega_0 + 2\omega_I)t$ is the mirror image of the original frequency ω_0.

$$\text{Original Frequency} \longrightarrow \omega_0$$

$$\text{Mirror Image Frequency} \longrightarrow \omega_0 + 2\omega_I$$

Mirror image frequency transmission stations must be avoided in the vicinity to avoid their reception.

4 Angle Modulation

Abstract

Building upon amplitude modulation, a detailed study of Angle Modulation is carried out in this chapter covering both Frequency Modulation (FM) and Phase Modulation (PM). The chapter begins by examining the fundamental principles of frequency and phase modulation, comparing peak versus average transmitted power levels, and analyzing the efficiency of different modulation techniques. Key practical aspects are introduced, such as Carson's Rule for estimating FM/PM bandwidth, Armstrong's method for generating wideband FM from narrowband FM, and receiver architectures designed to handle noise in real-world systems. A significant portion of the discussion focuses on noise analysis, contrasting the performance of AM and FM/PM systems. Concepts such as signal-to-noise ratio (SNR), noise modeling, and filtering techniques are systematically introduced, showing how FM and PM provide better protection against noise compared to AM. Both the advantages and disadvantages of FM/PM are highlighted, giving a balanced view of their role in communication systems.

U. Pillai and A. Patel, *Fundamentals of Communication Theory*,
https://doi.org/10.1007/978-3-032-20615-2_4

Table of Contents

1. Fourier Series

2. Fourier Transforms

3. Amplitude Modulation

4. Angle Modulation
4.1 Frequency and Phase Modulation
4.2 Peak vs. Average Transmitted Power
4.3 Better Protection against Noise in FM/PM
4.4 Disadvantages of FM/PM
4.5 Single Tone Angle Modulation
4.6 PM/FM Bandwidth: Carson's Rule
4.7 PM/FM Modulation Efficiency
4.8 Armstrong's Wideband FM Generation (from Narrowband FM)
4.9 FM receiver and Noise Analysis
4.10 SNR(Signal to Noise Power Ratio)
4.11 AM Noise Analysis
4.12 Noise Filtering
4.13 Noise Analysis in PM and FM

5. Digital Modulation Techniques

Frequency and Phase Modulation

In AM/DSB, message is accommodated in the amplitude of the carrier waveform $\cos\omega_0 t$, whereas in angle modulation the message is encoded in the phase or frequency of the carrier waveform.
Thus with $f(t)$ representing the message signal, in amplitude modulation (AM), the transmit signal $e(t)$ is given by:

$$e(t) = E_c\,(1 + \lambda f(t))\cos\omega_0 t \tag{4.1}$$

In angle modulation, the transmit signal $e(t)$ equals

$$e(t) = E_c \cos(\omega_0 t + \psi(t)) = E_c \cos\theta(t) \tag{4.2}$$

where the phase is modified by the message. In Phase Modulation (PM), the phase component $\psi(t)$ is proportional to the message. Thus

$$\psi(t) = k_p\, f(t) \tag{4.3}$$

Frequency and Phase Modulation

In Frequency Modulation (FM), the instantaneous frequency is proportional to the message. From (4.2), the instantaneous frequency $\omega(t)$ is given by

$$\omega(t) = \frac{d\theta(t)}{dt} = \frac{d}{dt}(\omega_0 t \;+\; \psi(t)) = \omega_0 t + \psi'(t), \tag{4.4}$$

and

$$\psi'(t) = k_f\, f(t). \tag{4.5}$$

Thus

$$\psi(t) = k_F \int_0^t f(\tau)\, d(\tau). \tag{4.6}$$

In summary in (4.2),

$$\psi(t) = \begin{cases} k_p\, f(t) & , \quad PM \\ k_F \int_0^t f(\tau)\, d(\tau) & , \quad FM \end{cases} \tag{4.7}$$

Frequency and Phase Modulation

Transmitted Signal

$$e(t) = E_c \cos(\omega_0 t + \psi(t)) = \begin{cases} E_c \cos(\omega_0 t + k_p f(t)) & , \quad PM \\ E_c \cos\left(\omega_0 t + k_F \int_0^t f(\tau)\, d(\tau)\right) & , \quad FM \end{cases} \tag{4.8}$$

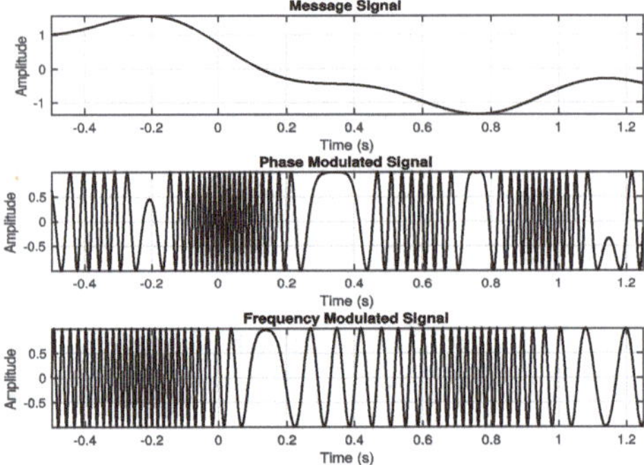

Figure 4.1: Phase and Frequency Modulation

Fourier Series Fourier Transforms Amplitude Modulation Angle Modulation Digital Modulation Techniques

Peak vs. Average Transmitted Power

From (4.8), peak transmitted power is $E_c^2/2$ since $|\cos\theta(t)| \leq 1$.
Also, average transmitted power is given by

$$\begin{aligned} \overline{e^2(t)} &= {E_c}^2 \frac{1}{T} \int_0^T \cos^2(\omega_0 t + \psi(t))\, dt \\ &= \frac{{E_c}^2}{2} \frac{1}{T} \int_0^T [1 + \cos(2\omega_0 t + 2\psi(t))]\, dt \\ &= \frac{{E_c}^2}{2} \end{aligned} \tag{4.9}$$

since the second terms at any significant frequency ω_0 averaged over a long duration goes to zero.
Hence, the peak power and average power in PM/FM are equal. This allows the transmitter to operate at saturation (peak efficiency) throughout the transmission. This is unlike inAM where the average transmit power is generally much lower than the peak power during most of the transmission duration.

Fundamentals of Communication Theory U. Pillai & A. Patel

Better Protection against Noise in FM/PM

Transmit signal in additive noise

$$\begin{aligned} r(t) &= e(t) + n(t) \\ &= E_c \cos(\omega_0 t + \psi(t)) + n(t) \\ &= E_c \cos\theta(t) + n(t) \quad ,\theta(t) = \omega_0 t + \psi(t) \\ &= [E_c + n(t)\cos\theta(t)]\cos\theta(t) + [n(t)\sin\theta(t)]\sin\theta(t) \\ &= A(t)\,\cos\left(\theta(t) + B(t)\right) \end{aligned} \tag{4.10}$$

where

$$\begin{aligned} A(t) &= \sqrt{(E_c + n(t)\cos\theta)^2 + (n(t)\sin\theta)^2} \\ &\simeq E_c + n(t)\cos\theta \end{aligned} \tag{4.11}$$

and

$$B(t) = \tan^{-1}\left(\frac{n(t)\sin\theta(t)}{E_c + n(t)\cos\theta(t)}\right) \simeq \frac{n(t)\sin\theta(t)}{E_c + n(t)\cos\theta(t)} \simeq \frac{n(t)\sin\theta(t)}{E_c} \tag{4.12}$$

when $E_c >> n(t)$

Better Protection against Noise in FM/PM

Substituting (4.11) - (4.12) in (4.10), we get the noisy return to be

$$r(t) = [E_c + n(t)\cos\theta(t)]\cos\left(\omega_0 t + \psi(t) + \frac{n(t)\sin\theta(t)}{E_c}\right) \tag{4.13}$$

Notice that the phase component of the noise is scaled by the original transmit signal amplitude E_c and hence its effect on the signal part $\psi(t)$ is much smaller compared to AM where the noise directly affects the message as can be seen from the amplitude part of (4.13).

This potential for FM/PM to significantly reduce the effect of noise is another advantage and will be discussed later in detail.

Disadvantages of FM/PM

In PM/FM, the message $f(t)$ is non-linearly modulated onto the carrier phase as in (4.8). To evaluate the bandwidth requirements expand (4.8) as

$$\begin{aligned} e(t) &= E_c \cos\left(\omega_0 t + k_p f(t)\right) \\ &= E_c \left(1 - \frac{[\omega_0 t + k_p f(t)]^2}{2!} + \frac{[\omega_0 t + k_p f(t)]^4}{4!} - \ldots + \ldots\right) \end{aligned} \tag{4.14}$$

(4.14) contains $f(t)$ as well as its higher powers $f^2(t), f^4(t), \ldots$ and hence once can easily see that it is a nonlinear function of $f(t)$.

Disadvantages of FM/PM

Thus with

$$f(t) \longleftrightarrow F(\omega) \quad , |\omega| \leq B_0$$

we have

$$\begin{aligned} f^2(t) &\longleftrightarrow \frac{1}{2\pi} F(\omega) \circledast F(\omega) \\ f^4(t) &\longleftrightarrow \frac{1}{2\pi} F(\omega) \circledast F(\omega) \circledast F(\omega) \circledast F(\omega) \end{aligned} \tag{4.15}$$

If $f(t)$ has bandwidth of $2B_0$, then the bandwidth of $f^2(t)$ is $4B_0$ and that of $f^4(t)$ is $8B_0$ etc. Thus from (4.14) - (4.15), PM/FM requires in principle infinite bandwidth.

Single Tone Angle Modulation

To understand PM/FM bandwidth implications in more detail, consider a single tone transmit signal

$$f(t) = E_0 \sin B_0 t \tag{4.16}$$

$$\psi(t) = k_p E_0 \sin B_0 t = \beta \sin B_0 t \tag{4.17}$$

where

$$\beta = k_p E_0 \tag{4.18}$$

represents the phase modulation index.
With (4.17) in (4.8),

$$\begin{aligned} e(t) &= \cos(\omega_0 t + \beta \sin B_0 t) = Re\left\{e^{j(\omega_0 t + \beta \sin B_0 t)}\right\} \\ &= Re\left\{e^{j\omega_0 t} \cdot e^{j\beta \sin B_0 t}\right\} = Re\left\{e^{j\omega_0 t} \cdot g(t)\right\} \end{aligned} \tag{4.19}$$

Fourier Series Fourier Transforms Amplitude Modulation Angle Modulation Digital Modulation Techniques

Single Tone Angle Modulation

where

$$g(t) = e^{j\beta \sin B_0 t} \tag{4.20}$$

is a periodic signal with period $T_0 = 2\pi/B_0$. Hence using Fourier's Theorem

$$g(t) = e^{j\beta \sin B_0 t} = \sum_{k=-\infty}^{\infty} c_k\, e^{jkB_0 t} \tag{4.21}$$

where

$$c_k = \frac{1}{T_0}\int_{-T_0/2}^{T_0/2} g(t)\, e^{-jkB_0 t}\, dt = \frac{B_0}{2\pi}\int_{-T_0/2}^{T_0/2} e^{-j(\beta \sin B_0 t - kB_0 t)}\, dt \tag{4.22}$$

Fundamentals of Communication Theory U. Pillai & A. Patel

Single Tone Angle Modulation

Let $B_0 t = \theta$, so that $dt = d\theta / B_0$ and (4.22) becomes

$$\begin{aligned} c_k &= \frac{B_0}{2\pi} \int_{-\pi}^{\pi} e^{j(\beta \sin\theta - k\theta)} \frac{d\theta}{B_0} = \frac{1}{2\pi} \int_{-\pi}^{\pi} e^{j(\beta \sin\theta - k\theta)}\, d\theta \\ &= \frac{1}{2\pi} \int_{-\pi}^{\pi} \cos(\beta \sin\theta - k\theta)\, d\theta \end{aligned} \tag{4.23}$$

since

$$\int_{-\pi}^{\pi} \sin(\beta \sin\theta - k\theta)\, d\theta = 0$$

Notice c_k in (4.23) are real and they represent the Bessel coefficients of order k.

Single Tone Angle Modulation

Hence

$$\begin{aligned} c_k = J_k(\beta) &= \frac{1}{2\pi} \int_{-\pi}^{\pi} \cos(\beta \sin x - kx)\, dx \\ &= \frac{1}{\pi} \int_{0}^{\pi} \cos(\beta \sin x - kx)\, dx \end{aligned} \tag{4.24}$$

Substituting (4.21) - (4.24) in (4.19), we get

$$\begin{aligned} e(t) &= E_c \cdot Re\left\{ e^{j\omega_0 t} \sum_{k=-\infty}^{\infty} J_k(\beta) e^{jkB_0 t} \right\} \\ &= E_c \sum_{k=-\infty}^{\infty} J_k(\beta) \cdot Re\left\{ e^{j(\omega_0 + kB_0)t} \right\} \\ &= E_c \sum_{k=-\infty}^{\infty} J_k(\beta) \cos(\omega_0 + kB_0)\, t \end{aligned} \tag{4.25}$$

Single Tone Angle Modulation

or

$$\begin{aligned} e(t) &= E_c \cos(\omega_0 t + \beta \sin B_0 t) \\ &= E_c \left[\ldots + J_{-k}(\beta)\cos(\omega_0 - kB_0)t + \ldots + J_{-1}(\beta)\cos(\omega_0 - B_0)t \right. \\ &\qquad \left. + J_0(\beta)\cos\omega_0 t + J_1(\beta)\cos(\omega_0 + B_0)t + \ldots \right] \end{aligned} \tag{4.26}$$

From (4.26), the angle modulated transmit signal contains infinite number of frequency components located at $\omega_0 + nB_0$, $n \in (-\infty, \infty)$ with associated power levels $E_c{}^2 J_n{}^2(\beta)/2$.

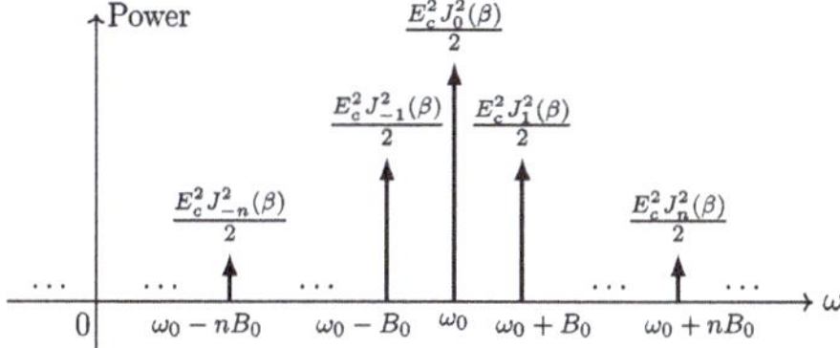

Figure 4.2: Power as a function of Bessel coefficients

Single Tone Angle Modulation

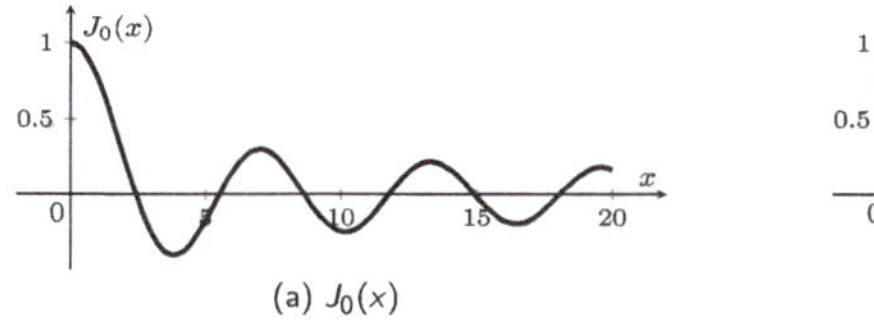

(a) $J_0(x)$

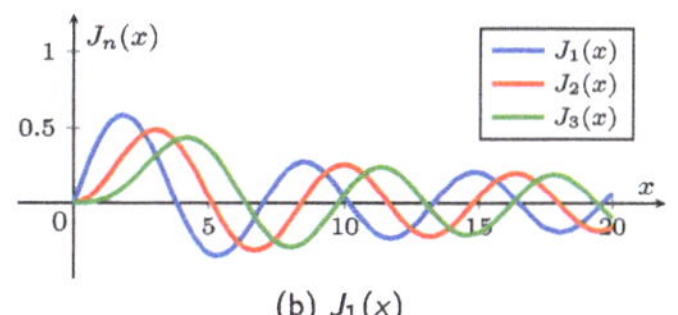

(b) $J_1(x)$

Figure 4.3: Bessel Functions $J_k(x)$, $k = 0, 1$

From (4.26), the average transmit power is given by

$$\begin{aligned} \overline{e^2(t)} &= E_c{}^2\, \overline{\cos^2(\omega_0 t + \beta \sin B_0 t)} = E_c{}^2/2 \\ &= E_c{}^2\, \overline{\left(\sum_{n=-\infty}^{\infty} J_k(\beta)\cos(\omega_0 + nB_0)t \right)^2} \\ &= E_c{}^2 \sum_{n=-\infty}^{\infty} J_n{}^2(\beta)\overline{\cos^2(\omega_0 + nB_0 t)} = E_c{}^2 \sum_{n=-\infty}^{\infty} J_n{}^2(\beta)\Big/2 \end{aligned} \tag{4.27}$$

Single Tone Angle Modulation

since

$$\overline{\cos(\omega_0 + kB_0)t \cdot \cos(\omega_0 + nB_0)t} = 0 \tag{4.28}$$

From (4.27), we get the identity

$$\sum_{n=-\infty}^{\infty} {J_n}^2(\beta) = 1 \tag{4.29}$$

or

$$J_0^2(\beta) + \sum_{n=1}^{\infty} \left[{J_n}^2(\beta) + {J_{-n}}^2(\beta)\right] = 1 \tag{4.30}$$

But from (4.24)

$$J_{-n}(\beta) = \frac{1}{2\pi}\int_{-\pi}^{\pi} \cos(\beta \sin x + nx)\, dx \tag{4.31}$$

Let $\theta = x + \pi$, so that

$$J_{-n}(\beta) = \frac{1}{2\pi}\int_{0}^{2\pi} \cos(-\beta \sin\theta + n(\theta - \pi))\, d\theta \tag{4.32}$$

Single Tone Angle Modulation

$$\begin{aligned} J_{-n}(\beta) &= \frac{1}{2\pi}\int_{0}^{2\pi} \cos(\beta \sin\theta - n\theta + n\pi)\, d\theta \\ &= (-1)^n \frac{1}{2\pi}\int_{0}^{2\pi} \cos(\beta \sin\theta - n\theta)\, d\theta = (-1)^n J_n(\beta) \end{aligned} \tag{4.33}$$

Using (4.33) in (4.30), we get the useful relation

$$J_0^2(\beta) + 2\sum_{n=1}^{\infty} {J_n}^2(\beta) = 1 \tag{4.34}$$

Thus

$$J_n(\beta) \to 0 \quad \text{as} \quad n \to \infty \quad , \quad \text{for any } \beta \tag{4.35}$$

We can use (4.34) - (4.35) to define an effective bandwidth.

PM/FM Bandwidth: Carson's Rule

Carson's Rule

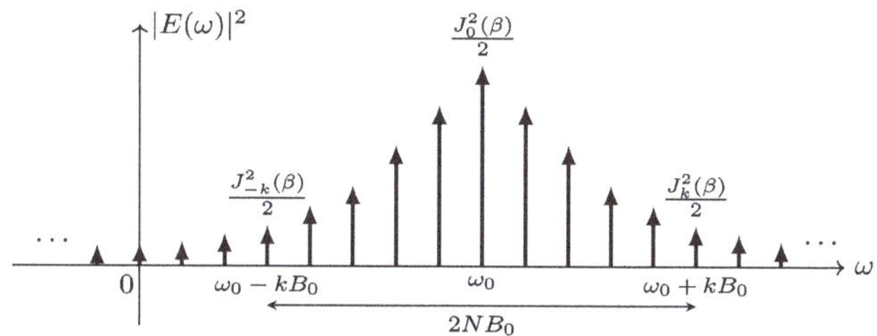

Figure 4.4: Effective Bandwidth in PM/FM

From (4.27) - (4.34) with $E_c = 1$

$$\overline{e^2(t)} = \frac{1}{2\pi}\overline{|E(\omega)|^2} = J_0{}^2(\beta) + 2\sum_{n=1}^{\infty} J_n{}^2(\beta) = 1 \tag{4.36}$$

i.e., power in the infinite number of frequency components adds up to unity (total power), and from (4.35), power in high frequency components go to zero.

Fourier Series Fourier Transforms Amplitude Modulation Angle Modulation Digital Modulation Techniques

PM/FM Bandwidth: Carson's Rule

We may select N such that 98% of energy in (4.36) is preserved, i.e., approximate (4.26) as

$$e(t) \simeq \sum_{n=-N}^{N} J_n(\beta)\cos(\omega_0 + nB_0)t \tag{4.37}$$

such that

$$J_0{}^2 + 2\sum_{n=1}^{N} J_n{}^2(\beta) \geq 0.98 \tag{4.38}$$

i.e., 98% of total power in (4.27) is preserved by the $2N+1$ components in (4.37).

Fundamentals of Communication Theory U. Pillai & A. Patel

PM/FM Bandwidth: Carson's Rule

Carson has emperically observed from Bessel function tables that so long as $\beta \leq 29$ integer N satisfying in (4.38) is given by

$$N \simeq \beta + 1 \tag{4.39}$$

i.e., if N is the nearest integer beyond $\beta + 1$, then the inequality in (4.38) is satisfied. Hence PM bandwidth

$$BW_{PM} = 2NB_0 \simeq 2(\beta + 1)B_0 \tag{4.40}$$

FM bandwidth

Message $f(t) = E_0 \cos B_0 t$

$$\psi(t) = k_F \int_0^t f(t) = \frac{k_F E_0}{B_0} \sin B_0 t = \beta \sin B_0 t \tag{4.41}$$

where

$$\beta = \frac{k_F E_0}{B_0} \tag{4.42}$$

represents the FM modulation index. Using (4.8) and (4.27)

$$e(t) = E_c \cos(\omega_0 t + \beta \sin B_0 t) = E_c \sum_{n=-\infty}^{\infty} J_n(\beta) \cos(\omega_0 + nB_0)t \tag{4.43}$$

From (4.37), Eq.(4.39) holds for FM also, and

$$\begin{aligned} BW_{FM} &= 2NB_0 = 2(\beta + 1)B_0 = 2\left(\frac{E_0 k_F}{B_0} + 1\right) B_0 \\ &= 2(E_0 k_F + B_0) = 2(\beta B_0 + B_0) \end{aligned} \tag{4.44}$$

Bandwidth Intervals in Frequency Deviation

From (4.43), instantaneous frequency in FM is given by

$$\omega = \frac{d}{dt}(\omega_0 t + \beta \sin B_0 t) = \omega_0 + \beta B_0 \cos \omega_0 t \qquad (4.45)$$

Maximum frequency deviation $\Delta\omega$ of FM carrier signals equals

$$\Delta\omega = \beta B_0 \qquad (4.46)$$

From (4.44)

$$BW_{FM} = 2(\Delta\omega + B_0) \qquad (4.47)$$

For PM, from (4.16) - (4.18), transmit signal equals

$$\begin{aligned} e(t) &= E_c \cos(\omega_0 t + k_P E_0 \sin B_0 t) \\ &= E_c \cos(\omega_0 t + \beta \sin B_0 t) \end{aligned} \qquad (4.48)$$

Bandwidth Intervals in Frequency Deviation

Frequency Deviation is given by

$$\frac{d}{dt}(\omega_0 t + \beta \sin B_0 t) = \omega_0 + \beta B_0 \cos B_0 t \qquad (4.49)$$

$$\Rightarrow \omega_0 - \beta B_0 \leq \omega \leq \omega_0 + \beta B_0 \Rightarrow \Delta\omega = \beta B_0 \qquad (4.50)$$

Narrowband PM/FM $\Rightarrow \beta \simeq 0$
From (4.44)

$$BW = 2B_0 \qquad (4.51)$$

Carson's rule in (4.40) - (4.44) usually underestimates Wideband PM/FM bandwidth if $\beta > 3$. In those cases

$$BW = 2(\beta + 4)B_0 = 2(\Delta\omega + 4B_0) \qquad (4.52)$$

is more accurate for BW and may be chosen over (4.47)

PM/FM Modulation Efficiency

From (4.25) - (4.27), $E_c\, J_0(\beta)$ represents the carrier component and $E_c\, {J_0}^2(\beta)/2$ represents the transmit power associated with the carrier component (see Fig (4.3)). Also, ${E_c}^2/2$ represents the total power.

Hence, $\frac{{E_c}^2}{2}\left(1 - {J_0}^2(\beta)\right)$ represents the power associated with the message bearing components.

Thus, the index given by

$$\eta = 1 - {J_0}^2(\beta) \tag{4.53}$$

represents the efficiency of angle modulation.

$$\Rightarrow \text{If } J_0(\beta_0) = 0 \Rightarrow \eta = 1$$

Table below gives the Bessel function values.

Bessel Function Table

$\beta \backslash n$	0	1	2	3	4	5	6	7	8	9	10
0.0	1.0000	0.0000	0.0000	0.0000	0.0000	0.0000	0.0000	0.0000	0.0000	0.0000	0.0000
0.1	0.9975	0.0499	0.0012	0.0000	0.0000	0.0000	0.0000	0.0000	0.0000	0.0000	0.0000
0.2	0.9900	0.0995	0.0050	0.0002	0.0000	0.0000	0.0000	0.0000	0.0000	0.0000	0.0000
0.3	0.9776	0.1483	0.0112	0.0006	0.0000	0.0000	0.0000	0.0000	0.0000	0.0000	0.0000
0.4	0.9604	0.1960	0.0197	0.0013	0.0001	0.0000	0.0000	0.0000	0.0000	0.0000	0.0000
0.5	0.9385	0.2423	0.0306	0.0026	0.0002	0.0000	0.0000	0.0000	0.0000	0.0000	0.0000
0.6	0.9120	0.2867	0.0437	0.0044	0.0003	0.0000	0.0000	0.0000	0.0000	0.0000	0.0000
0.7	0.8812	0.3290	0.0588	0.0069	0.0006	0.0000	0.0000	0.0000	0.0000	0.0000	0.0000
0.8	0.8463	0.3688	0.0758	0.0102	0.0010	0.0001	0.0000	0.0000	0.0000	0.0000	0.0000
0.9	0.8075	0.4059	0.0946	0.0144	0.0016	0.0001	0.0000	0.0000	0.0000	0.0000	0.0000
1.0	0.7652	0.4401	0.1149	0.0196	0.0025	0.0002	0.0000	0.0000	0.0000	0.0000	0.0000
1.5	0.5118	0.5579	0.2321	0.0610	0.0118	0.0018	0.0002	0.0000	0.0000	0.0000	0.0000
2.0	0.2239	0.5767	0.3528	0.1289	0.0340	0.0070	0.0012	0.0002	0.0000	0.0000	0.0000
2.5	-0.0484	0.4971	0.4461	0.2166	0.0738	0.0195	0.0042	0.0008	0.0001	0.0000	0.0000
3.0	-0.2601	0.3391	0.4861	0.3091	0.1320	0.0430	0.0114	0.0025	0.0005	0.0001	0.0000
3.5	-0.3801	0.1374	0.4586	0.3868	0.2044	0.0804	0.0254	0.0067	0.0015	0.0003	0.0001
4.0	-0.3971	-0.0660	0.3641	0.4302	0.2811	0.1321	0.0491	0.0152	0.0040	0.0009	0.0002
4.5	-0.3097	-0.2201	0.2782	0.4247	0.3484	0.1947	0.0843	0.0300	0.0091	0.0024	0.0006
5.0	-0.1776	-0.3276	0.1085	0.3648	0.3912	0.2611	0.1310	0.0534	0.0184	0.0055	0.0015

Table 4.1: Bessel function values $J_n(\beta)$.

Armstrong's Wideband FM Generation (from Narrowband FM)

For FM, carrier waveform is given by

$$e(t) = \cos(\omega_0 t + \psi(t)) \tag{4.54}$$

where

$$\psi(t) = k_F \int_0^T f(\tau)d\tau = \beta g(t) \tag{4.55}$$

and β is the modulation index. Thus transmit waveform equals

$$e(t) = \cos(\omega_0 t + \beta g(t)) \tag{4.56}$$

When β is small, then

$$\begin{aligned} e(t) &= \cos\omega_0 t \cos(\beta g(t)) - \sin(\beta g(t)) \sin\omega_0 t \\ &\simeq \cos\omega_0 t - \beta g(t) \sin\omega_0 t \end{aligned} \tag{4.57}$$

which is essentially amplitude modulation (AM) requiring $2B_0$ bandwidth, where B_0 represents the bandwidth of the message signal.

Armstrong's Wideband FM Generation (from Narrowband FM)

Since the bandwidth requirement for FM is low when β is small, it is known as Narrowband FM. Implementation of Narrowband FM is same as AM as shown below

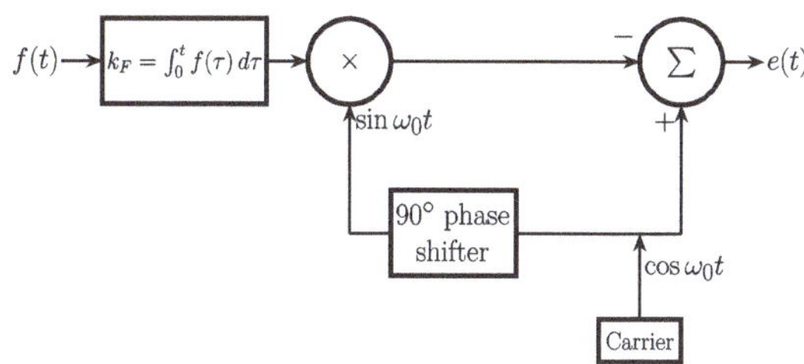

Figure 4.5: Narrowband FM generation

Armstrong's Wideband FM Generation (from Narrowband FM)

Armstrong's "genius idea" was in figuring out a way to convert a Narrowband FM (β small) to a Wideband FM (β large). He realized this using a frequency multiplier device consisting of a square law device (non-linear device) followed by a band pass filter as shown in Fig(4.6), and repeating this module multiple times.

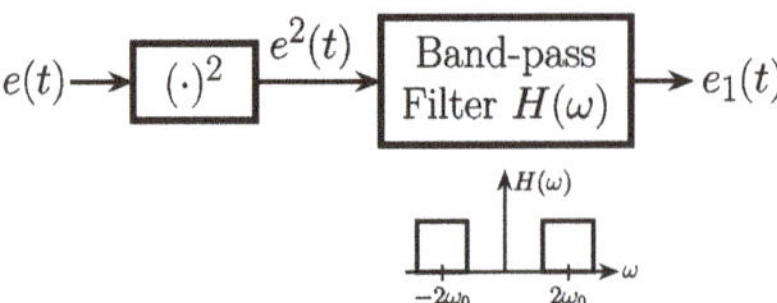

Figure 4.6: Narrowband FM generation

Fourier Series Fourier Transforms Amplitude Modulation Angle Modulation Digital Modulation Techniques

Armstrong's Wideband FM Generation (from Narrowband FM)

$$e^2(t) = \cos^2(\omega_0 t + \beta g(t)) = \frac{1 + \cos(2\omega_0 t + 2\beta g(t))}{2} \tag{4.58}$$

so that after band-pass around center frequency $2\omega_0$, we get

$$e_1(t) = \frac{1}{2}\cos(2\omega_0 t + 2\beta g(t)). \tag{4.59}$$

Eq(4.59) is similar to (4.58), except the modulation index has increased to 2β from β, along with doubling of carrier frequency to $2\omega_0$. Carrier frequency can be demodulated to ω_0 as in Fig. 4.7

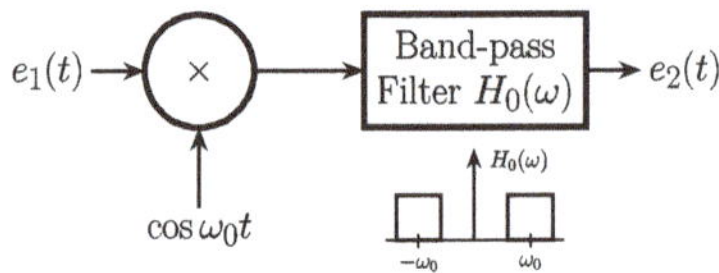

Figure 4.7: Narrowband FM generation

Armstrong's Wideband FM Generation (from Narrowband FM)

Using (4.59)

$$\begin{aligned} e_1(t)\cos\omega_0 t &= \frac{1}{2}\cos(2\omega_0 t + 2\beta g(t))\cos\omega_0 t \\ &= \frac{1}{4}\cos(\omega_0 t + 2\beta g(t)) + \frac{1}{4}\cos(3\omega_0 t + 2\beta g(t)) \end{aligned} \tag{4.60}$$

Band-pass filter $H_0(\omega)$ is used to suppress the second term in (4.60), so that its output is equals

$$e_2(t) = \frac{1}{4}\cos(\omega_0 t + 2\beta g(t)) \tag{4.61}$$

Repeating this procedure, the modulation can be made to $2^k\beta$ after k such steps, generating wideband FM.

FM receiver and Noise Analysis

FM transmit signal

$$e(t) = E_c\cos(\omega_0 t + \beta g(t)) \tag{4.62}$$

$$g(t) = \int_0^t f(\tau)\,d\tau \tag{4.63}$$

To recover the message $f(t)$ riding in the phase component, we can use a differentiator followed by an envelope detector.

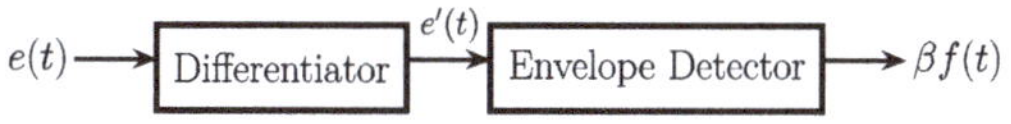

$e(t)$ → Differentiator → $e'(t)$ → Envelope Detector → $\beta f(t)$

Figure 4.8: FM Receiver

The differentiator output gives

$$\begin{aligned} e'(t) &= -E_c(\omega_0 + \beta g'(t))\sin(\omega_0 + \beta g(t)) \\ &= -E_c(\omega_0 + \beta f(t))\sin(\omega_0 + \beta g(t)) \end{aligned} \tag{4.64}$$

and the envelope detector outputs the desired message signal $f(t)$.

FM receiver and Noise Analysis

However, in reality $e(t)$ will be contaminated with noise $n(t)$ as

$$r(t) = e(t) + n(t) \tag{4.65}$$

where $n(t)$ can be modeled as random noise with frequency contributions over a wide spectral region. Thus

$$n(t) = \sum_k n_k(t) = \sum_k A_k \cos(\omega_k t + \theta_k) \tag{4.66}$$

where ω_k indicates the associated frequency and θ_k is a random variable that is uniformly distributed in $(0, 2\pi)$ i.e., $\theta_k \sim U(0, 2\pi)$.

With (4.66) in (4.65), the received signal has the form

$$\begin{aligned} r(t) &= e(t) + n(t) \\ &= E_c \cos(\omega_0 t + \psi(t)) + \sum_k A_k \cos(\omega_k + \theta_k) \end{aligned} \tag{4.67}$$

where each of the individual noise components $n_k(t)$ are weak compared to the signal component.

FM receiver and Noise Analysis

To rewrite (4.67) as a single component $A(t)\cos\psi(t)$, we can make use of the Alignment Rule.

Alignment Rule

Consider the sum of n real signals

$$e(t) = a_1(t)\cos\psi_1(t) + a_2(t)\cos\psi_2(t) + \ldots + a_n(t)\cos\psi_n(t) \tag{4.68}$$

where the instantaneous frequency of each component signal is given by $\frac{d\psi_1(t)}{dt}$. To determine the instantaneous frequency of the combined signal $e(t)$, we need to rewrite it in the form

$$e(t) = A(t)\cos\psi(t) \tag{4.69}$$

so that

$$\omega(t) = \frac{d\psi(t)}{dt} \tag{4.70}$$

represents the instantaneous frequency of the combined signal in (4.68).

FM receiver and Noise Analysis

Towards this, rewrite the sum in (4.68) as

$$\begin{aligned}
e(t) &= \sum_{k=1}^{n} a_k \cos\psi_k(t) = Re\left\{\sum_{k=1}^{n} a_k(t)e^{j\psi_k(t)}\right\} \\
&= Re\left\{a_1(t)e^{j\psi_1(t)} + \sum_{k=2}^{n} a_k(t)e^{j\psi_k(t)}\right\} \\
&= Re\left\{e^{j\psi_1(t)}\left(a_1(t) + \sum_{k=2}^{n} a_k(t)e^{j(\psi_k(t)-\psi_1(t))}\right)\right\} \\
&= Re\left\{e^{j\psi_1(t)}\left[a_1(t) + \sum_{k=2}^{n} a_k(t)\cos(\psi_k(t)-\psi_1(t))\right.\right. \\
&\qquad \left.\left. + j\sum_{k=2}^{n} a_k(t)\sin(\psi_k(t)-\psi_1(t))\right]\right\} \\
&= Re\left\{e^{j\psi_1(t)}\left[A_1(t) + jA_2(t)\right]\right\}
\end{aligned} \tag{4.71}$$

Fourier Series Fourier Transforms Amplitude Modulation Angle Modulation Digital Modulation Techniques

FM receiver and Noise Analysis

where

$$A_1(t) = a_1(t) + \sum_{k=2}^{n} a_k(t)\cos(\psi_k(t)-\psi_1(t)) \tag{4.72}$$

and

$$A_2(t) = \sum_{k=2}^{n} a_k(t)\sin(\psi_k(t)-\psi_1(t)) \tag{4.73}$$

Let

$$A_1(t) + jA_2(t) = A(t)e^{j\phi(t)} \tag{4.74}$$

so that

$$A(t) = \sqrt{{A_1}^2(t) + {A_2}^2(t)} \tag{4.75}$$

and

$$\phi(t) = \tan^{-1}\left(\frac{A_2(t)}{A_1(t)}\right) \tag{4.76}$$

Fundamentals of Communication Theory U. Pillai & A. Patel

FM receiver and Noise Analysis

Using (4.74) in (4.71), we get

$$\begin{aligned} e(t) &= Re\left\{e^{j\psi_1(t)}A(t)e^{j\phi(t)}\right\} = Re\left\{A(t)e^{j(\psi_1(t)+\phi(t))}\right\} \\ &= A(t)\cos(\psi_1(t)+\phi(t)) = A(t)\cos\psi(t) \end{aligned} \tag{4.77}$$

On comparing (4.77) to (4.69), $A(t)$ in (4.77) represents the overall envelope and

$$\psi(t) = \psi_1(t) + \phi(t) \tag{4.78}$$

represents the overall phase and

$$\omega(t) = \frac{d\psi(t)}{dt} = \psi_1'(t) + \phi'(t) \tag{4.79}$$

the overall instantaneous frequency of the combined signal $e(t)$ in (4.68).

Fourier Series Fourier Transforms Amplitude Modulation Angle Modulation Digital Modulation Techniques

FM receiver and Noise Analysis

If the signal component $a_1(t)$ is predominant in (4.68), i.e., if

$$|a_1(t)| >> |a_k(t)| \quad , \quad k = 2, 3, \dots n \tag{4.80}$$

then from (4.72) - (4.74), $A_1{}^2(t) >> A_2{}^2(t)$ and hence from (4.75)

$$A(t) \simeq A_1(t) = a_1(t) + \sum_{k=2}^{n} a_k(t)\cos(\psi_k(t) - \psi_1(t)) \tag{4.81}$$

and from (4.77)

$$\phi(t) \simeq \frac{A_2(t)}{A_1(t)} \simeq \frac{A_2(t)}{a_1(t)} = \frac{\sum_{k=2}^{n} a_k(t)\sin(\psi_k(t) - \psi_1(t))}{a_1(t)} \tag{4.82}$$

In summary if the dominant signal $|a_1(t)| >> |a_k(t)|, k = 2 \to n$, then

$$\begin{aligned} a_1(t)\cos\psi_1(t) + \sum_{k=2}^{n} a_k(t)\cos\psi_k(t) &\simeq \left[a_1(t) + \sum_{k=2}^{n} a_k(t)\cos(\psi_k(t) - \psi_1(t))\right] \\ &\cos\left[\psi_1(t) + \frac{\sum_{k=2}^{n} a_k(t)\sin(\psi_k(t) - \psi_1(t))}{a_1(t)}\right] \end{aligned} \tag{4.83}$$

Fundamentals of Communication Theory U. Pillai & A. Patel

FM receiver and Noise Analysis

Hence

$$m(t) = \sum_{k=2}^{n} a_k(t) \cos(\psi_k(t) - \psi_1(t)) \tag{4.84}$$

represents the amplitude "noise" distortion to $a_1(t)$ due to other components and

$$\theta(t) = \frac{\sum_{k=2}^{n} a_k(t) \sin(\psi_k(t) - \psi_1(t))}{a_1(t)} \tag{4.85}$$

represents the phase "noise" distortion to the original $\psi_1(t)$ due to other components.

Example: The frequency modulated (FM) signal $\cos(\omega_0 t + \beta_1 \sin B_1 t)$ is contaminated by another FM signal $\rho \cos(\omega_0 t + \beta_2 \sin B_2 t)$ where $|\rho| << 1$. Find the overall instantaneous frequency of the combined signal.

FM receiver and Noise Analysis

Solution

$$\begin{aligned} e(t) &= \cos(\omega_0 t + \beta_1 \sin B_1 t) + \rho \cos(\omega_0 t + \beta_2 \sin B_2 t) \\ &= A(t) \cos \psi(t) \end{aligned} \tag{4.86}$$

where using (4.83) - (4.85)

$$A(t) = 1 + \rho \cos(\beta_1 \sin B_1 t - \beta_2 \sin B_2 t) \tag{4.87}$$

and

$$\psi(t) = \omega_0 t + \beta_1 \sin B_1(t) + \rho \sin(\beta_2 \sin B_2 t - \beta_1 \sin B_1 t) \tag{4.88}$$

and the instantaneous frequency $\omega(t)$ is given by

$$\begin{aligned} \omega(t) = \frac{d\psi(t)}{dt} &= \omega_0 + \beta_1 B_1 \cos B_1 t + \rho(\beta_2 B_2 \cos B_2 t - \beta_1 B_1 \cos B_2 t) \\ &\qquad \cos(\beta_2 \sin B_2 t - \beta_1 \sin B_1 t) \end{aligned} \tag{4.89}$$

Next, we will use (4.83) - (4.85) to perform noise analysis in AM and PM/FM.

SNR(Signal to Noise Power Ratio)

In the presence of noise, the ratio of the signal power to the average noise power is taken as a performance measure. Thus with

$$e(t) = s(t) + n(t) \tag{4.90}$$

the quantity

$$\text{SNR} = \frac{\overline{s^2(t)}}{E[n^2(t)]} = \frac{E}{\sigma^2} \tag{4.91}$$

represents the SNR. Generally, SNR is computed both at the input and the output of the receiver to estimate the receiver performance.

$s(t) + n(t)$ → Receiver → $\hat{s}(t) + w(t)$

Figure 4.9: Noise Analysis

Fourier Series Fourier Transforms Amplitude Modulation Angle Modulation Digital Modulation Techniques

SNR(Signal to Noise Power Ratio)

Input SNR is given by

$$SNR_i = \frac{\overline{s^2(t)}}{E[n^2(t)]} \tag{4.92}$$

and output SNR

$$SNR_o = \frac{\overline{\hat{s}^2(t)}}{E[w^2(t)]} \tag{4.93}$$

and

$$\eta = \frac{SNR_o}{SNR_i} \tag{4.94}$$

may be chosen to represent the overall efficiency of the underlying modulation and receiver scheme to noise suppression.

Fundamentals of Communication Theory U. Pillai & A. Patel

AM Noise Analysis

Transmit AM signal

$$e_0(t) = E_0(1 + \lambda f(t)) \cos \omega_0 t \tag{4.95}$$

Figure 4.10: Noise Analysis in AM

Suppose the channel adds noise $n(t) \cos \omega_1 t$ so that the noisy received signal $e(t)$ is given by

$$e(t) = E_0(1 + \lambda f(t)) \cos \omega_0 t + n(t) \cos(\omega_1 t + \theta) \tag{4.96}$$

and using (4.83) - (4.85)

$$\begin{aligned} e(t) = [E_0 \, (1 + \lambda f(t) + n(t) \cos(\omega_1 - \omega_0)t + \theta)] \\ \cos\left(\omega_0 t + \frac{n(t) \sin(\omega_1 - \omega_0)t + \theta}{E_0(1 + \lambda f(t))}\right) \end{aligned} \tag{4.97}$$

AM Noise Analysis

From (4.92), the noise envelope $A(t)$ equals

$$A(t) = E_0(1 + \lambda f(t)) + n(t) \cos((\omega_1 - \omega_0)t + \theta) \tag{4.98}$$

and $\lambda f(t)$ represents the recovered signal part and $n(t) \cos((\omega_1 - \omega_0)t + \theta)$ the noise component in the recovered signal. Using (4.92) in (4.96), we get the input SNR to be

$$\begin{aligned} SNR_i &= \frac{{E_0}^2 \lambda^2 \overline{f^2(t)\cos^2 \omega_0 t}}{E[n^2(t) \cos^2(\omega_1 t + \theta)]} = \frac{{E_0}^2 \lambda^2 \overline{f^2(t)}/2}{E[n^2(t) \cos^2(\omega_1 t + \theta)]} \\ &= \frac{{E_0}^2 \lambda^2 \overline{f^2(t)}/2}{E[n^2(t)]/2} = \frac{{E_0}^2 \lambda^2 \overline{f^2(t)}}{E[n^2(t)]} \end{aligned} \tag{4.99}$$

and from (4.93) - (4.97), the output SNR to be

$$SNR_o = \frac{{E_0}^2 \lambda^2 \overline{f^2(t)}}{E[n^2(t) \cos^2((\omega_1 - \omega_0)t + \theta)]} \simeq \frac{{E_0}^2 \lambda^2 \overline{f^2(t)}}{E[n^2(t)]/2} \tag{4.100}$$

AM Noise Analysis

so that

$$\eta = \frac{SNR_o}{SNR_i} = \frac{2{E_0}^2\lambda^2\overline{f^2(t)}}{E[n^2(t)]} \cdot \frac{E[n^2(t)]}{{E_0}^2\lambda^2\overline{f^2(t)}} = 2 \tag{4.101}$$

i.e.,
For AM, the improvement in SNR is only 2 (or 3dB) from the input to the output of the envelope detector and it is independent of the transmit signal amplitude.

Noise Modelling

The simplest model for noise is to represent it as

$$n(t) = \sum_k n_k(t) = \sum_k A_k \cos(\omega_k t + \theta_k) \tag{4.102}$$

where ω_k indicates the associate frequency and θ_k is a random phase variable that is uniformly distributed in $(0, 2\pi)$, i.e., $\theta_k \sim U(0, 2\pi)$, where θ_ks are independent across different frequencies. Thus

$$\begin{aligned} E[n_k(t)] &= E[A_k \cos(\omega_k t + \theta_k)] \\ &= E[A_k(\cos\omega_k t \cos\theta_k - \sin\omega_k t \sin\theta_k)] \\ &= A_k[\cos\omega_k t\, E(\cos\theta_k) - \sin\omega_k t\, E(\sin\theta_k)] \\ &= 0 \end{aligned} \tag{4.103}$$

since

$$E(\cos\theta_k) = \int_0^{2\pi} \cos\theta_k \cdot f_{\theta_k}(\theta_k)\, d\theta_k = \int_0^{2\pi} \cos\theta \cdot \frac{1}{2\pi}\, d\theta = 0 \tag{4.104}$$

Noise Modeling

Similarly,

$$\begin{aligned} E[n_k(t)n_j(t)] &= A_k A_j\, E[\cos(\omega_k t + \theta_k)\cos(\omega_j t + \theta_j)] \\ &= A_k A_j\, E[\cos(\omega_k t + \theta_k)]\, E[\cos(\omega_j t + \theta_j)] = 0 \end{aligned} \tag{4.105}$$

since θ_k and θ_j are independent random variables for $i \neq j$. Moreover the average power at frequency ω_k is given by

$$\begin{aligned} P_k = E[n^2(t)] &= {A_k}^2\, E[\cos^2(\omega_k t + \theta_k)] \\ &= \frac{{A_k}^2}{2}\, E[1 + \cos(2\omega_k t + 2\theta_k)] \\ &= \frac{{A_k}^2}{2} + \frac{{A_k}^2}{2}\, \underbrace{E[\cos(2\omega_k t + 2\theta_k)]}_{0} = \frac{{A_k}^2}{2} \end{aligned} \tag{4.106}$$

Fourier Series Fourier Transforms Amplitude Modulation **Angle Modulation** Digital Modulation Techniques

Noise Modeling

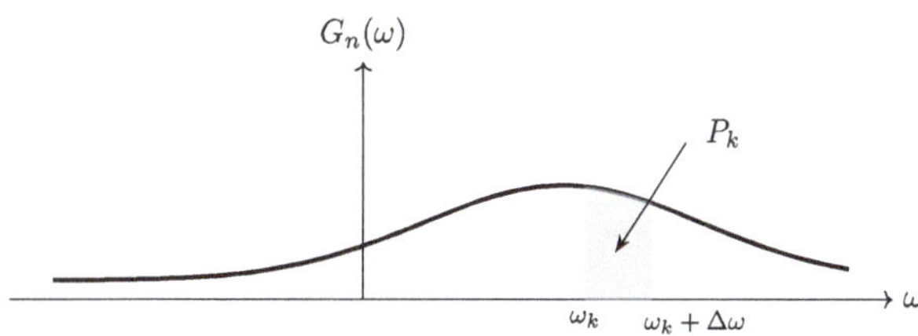

Figure 4.11: Noise Power Spectral Density vs. Frequency

Thus, P_k represents the noise power in the frequency bin $(\omega_k, \omega_k + \Delta\omega)$ and if we define

$$G(\omega_k) = \frac{P_k}{\Delta\omega_k} = \frac{{A_k}^2/2}{\Delta\omega} \tag{4.107}$$

then $G(\omega_k)$ defines the power density of the noise signal at frequency $\omega = \omega_k$.

Fundamentals of Communication Theory U. Pillai & A. Patel

Noise Modeling

The total power contained in the noise signal $n(t)$ in (4.102) equals

$$\begin{aligned} P = E[n^2(t)] = E\left[\left(\sum_k n_k(t)\right)^2\right] &= E\left[\sum_k n_k(t) \sum_j n_j(t)\right] \\ &= \sum_k E[{n_k}^2(t)] + \overbrace{\sum_k \sum_{\substack{j \\ k \neq j}} E[n_k(t)\, n_j(t)]}^{0} \\ &= \sum_k E[{n_k}^2(t)] = \sum_k P_k = \sum_k G_n(\omega_k)\, \Delta\omega \\ &= \int_{-\infty}^{\infty} G_n(\omega)\, d\omega \end{aligned} \tag{4.108}$$

If the noise has equal power at all frequencies, then

$$G_n(\omega) = \sigma^2 \quad , \quad -\infty < \omega < +\omega \tag{4.109}$$

Fourier Series Fourier Transforms Amplitude Modulation Angle Modulation Digital Modulation Techniques

Noise Modeling

and such a noise process is known as white noise process (similar to sun light that contains all visible frequencies more or less at equal strength.)

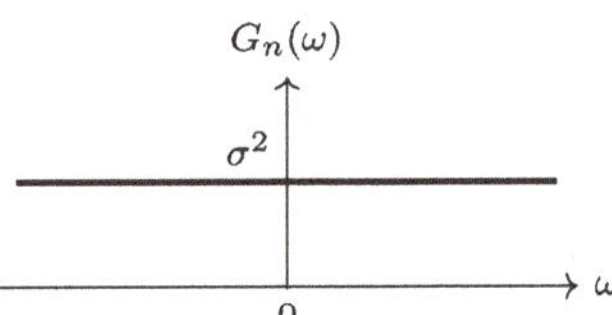

Figure 4.12: White noise power spectrum

In that case, from (4.106) - (4.109)

$$P_k = {A_k}^2/2 = \sigma^2\, \Delta\omega \tag{4.110}$$

so that $A_k = 2\sigma^2\, \Delta\omega$ is a constant and (4.102) can be rewritten as

$$n(t) = A \sum_k \cos(\omega_k t + \theta_k) \tag{4.111}$$

Noise Filtering

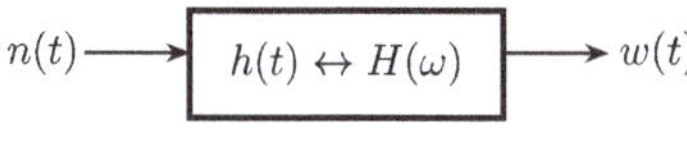

Figure 4.13: Noise Filtering

Suppose $n(t) = \sum A_k \cos(\omega_k t + \theta_k)$ is passed through a linear time-invariant filter with impulse response $h(t)$ to generate output noise $w(t)$. Thus

$$w(t) = \sum_k w_k(t)$$

where

Noise Filtering

$$\begin{aligned}
w_k(t) &= \mathcal{L}\{A_k \cos(\omega_k t + \theta_k)\} = A_k \cos(\omega_k t + \theta_k) \circledast h(t) \\
&= Re\left\{A_k e^{+j(\omega_k t + \theta_k)} \circledast h(t)\right\} = Re\left\{A_k \int_{-\infty}^{+\infty} e^{j(\omega_k(t-\tau)+\theta_k)} h(\tau)\, d\tau\right\} \\
&= Re\left\{ A_k e^{j(\omega_k t + \theta_k)} \underbrace{\int_{-\infty}^{+\infty} h(\tau) e^{-j\omega_k \tau}\, d\tau}_{H(\omega_k) = |H(\omega_k)| e^{j\psi_k(\omega)}} \right\} \\
&= Re\left\{A_k |H(\omega_k)| e^{j(\omega_k t + \theta_k + \psi_k(\omega))}\right\} \\
&= A_k |H(\omega_k)|^2 \cos(\omega_k t + \theta_k + \psi_k(\omega)) \\
&= B_k \cos(\omega_k t + \theta_k + \psi_k(\omega))
\end{aligned} \tag{4.112}$$

Noise Filtering

On comparing with (4.106), the output power at frequency $\omega = \omega_k$

$$E({\omega_k}^2(t)) = \frac{{B_k}^2}{2} = \left(\frac{{A_k}^2}{2}\right)|H(\omega_k)|^2 = G_w(\omega_k) \quad (4.113)$$

Thus from (4.107)

$$G_w(\omega_k) = G_n(\omega_k)|H(\omega_k)|^2 \quad (4.114)$$

or

$$G_w(\omega) = G_n(\omega)|H(\omega)|^2 \quad (4.115)$$

represents the input-output power spectral density relation for a random process applied to a filter with transfer function $h(t) \leftrightarrow H(\omega)$.

Fourier Series Fourier Transforms Amplitude Modulation Angle Modulation Digital Modulation Techniques

Noise Analysis in PM and FM

In both PM and FM, the transmit signal $e(t)$ has a large bandwidth compared to the message $f(t)$.

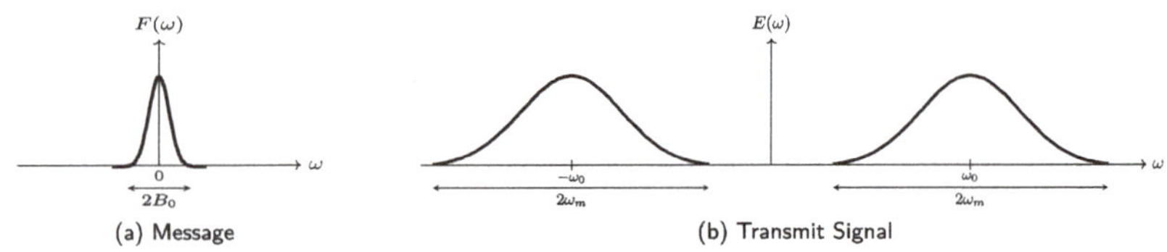

(a) Message (b) Transmit Signal

Figure 4.14: Message vs actual transmit signal

Received Signal

$$r(t) = e(t) + n(t) \quad (4.116)$$

where $n(t)$ is white noise of spectral density σ^2 as in Fig. 4.12

Fundamentals of Communication Theory U. Pillai & A. Patel

Noise Analysis in PM and FM

Initially the received signal, $r(t)$ is passed through a band-pass filter $H_1(\omega)$ (pre-filter) with bandwidth $2\omega_m$ centered at ω_0 to limit the noise as shown in Fig. 4.15.

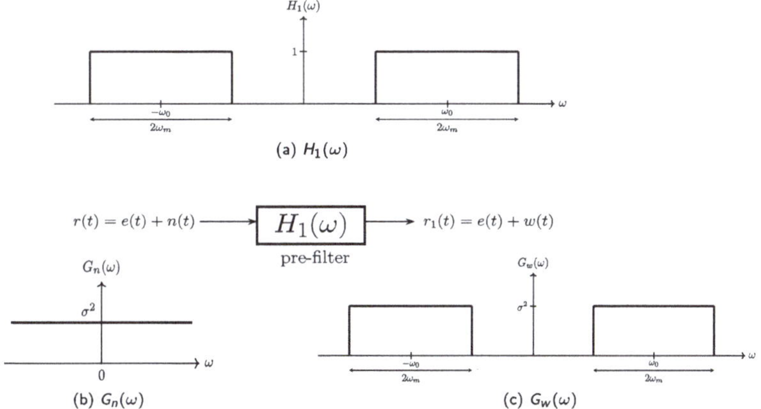

Figure 4.15: Pre-filter signal transformation

Fourier Series Fourier Transforms Amplitude Modulation Angle Modulation Digital Modulation Techniques

Noise Analysis in PM and FM

The pre-filtered signal $r_1(t)$ is then passed through a demodulator to generate the noisy message $f(t)$ and finally the output $g(t)$ is passed through a post-filter $H_2(\omega)$ to further limit the output noise to that of the recovered signal $f(t)$.

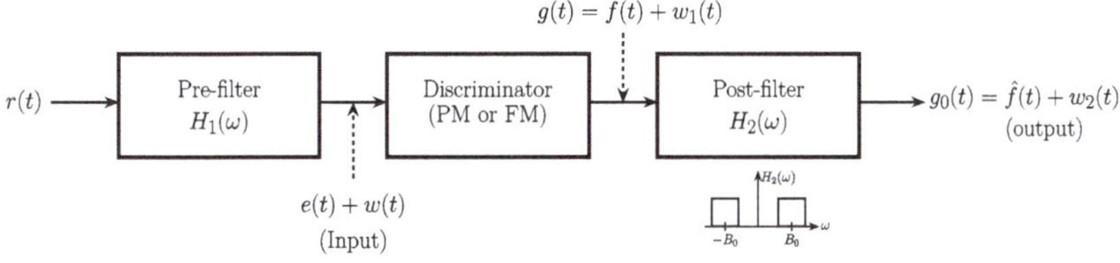

Figure 4.16: Block diagram

$$SNR_i = \frac{\text{Input signal power}}{\text{Average noise power}} = \frac{\overline{e^2(t)}}{E[w^2(t)]} \tag{4.117}$$

$$\overline{e^2(t)} = \overline{{E_c}^2 \cos^2(\omega_0 t + \psi(t))} = {E_c}^2/2$$

Noise Analysis in PM and FM

From Fig. 4.15c,

$$\begin{aligned} E[w^2(t)] &= \int_{-\infty}^{+\infty} G_w(\omega)\, d\omega = 2\int_{\omega_0-\omega_m}^{\omega_0+\omega_m} \sigma^2\, d\omega \\ &= 4\sigma^2\omega_m = 4\sigma^2\beta B_0 \end{aligned} \tag{4.118}$$

using Carson's Rule ($\omega_m \simeq \beta B_0$).
Hence

$$SNR_i = \frac{\overline{e^2(t)}}{E[w^2(t)]} = \frac{E_c{}^2}{8\sigma^2\beta B_0} \tag{4.119}$$

From (4.16), the discriminator Input signal

$$\begin{aligned} r_1(t) &= e(t) + w(t) \\ &= E_c\cos(\omega_0 t + \psi(t)) + \sum_k B_k\cos(\omega_k t + \theta_k) \end{aligned} \tag{4.120}$$

where the summation covers all frequencies in Fig. 4.15c.

Noise Analysis in PM and FM

Using alignment rule in (4.84) - (4.86)

$$\begin{aligned} r_1(t) &= \left[E_c + \sum B_k\cos((\omega_k-\omega_0)t + \theta_k - \psi(t))\right] \\ &\quad \cos\left(\omega_0 t + \psi(t) + \frac{\sum B_k\cos((\omega_k-\omega_0)t+\theta_k-\psi(t))}{E_c}\right) \\ &= A(t)\cos\psi_0(t) \end{aligned} \tag{4.121}$$

where

$$\psi_0(t) = \omega_0 t + \psi(t) + w_1(t)/E_c \tag{4.122}$$

where

$$w_1(t) = \sum_k B_k\cos((\omega_k-\omega_0)t + \theta_k - \psi(t)) \tag{4.123}$$

Noise Analysis in PM and FM

From (4.121), frequencies in $w_1(t)$ at the discriminator output in Fig. 4.16 are downshifted version of those in $w(t)$ by $\pm\omega_0$ with no change in amplitude. Thus from Fig. 4.15c, we obtain $G_{w_1}(\omega)$ to be

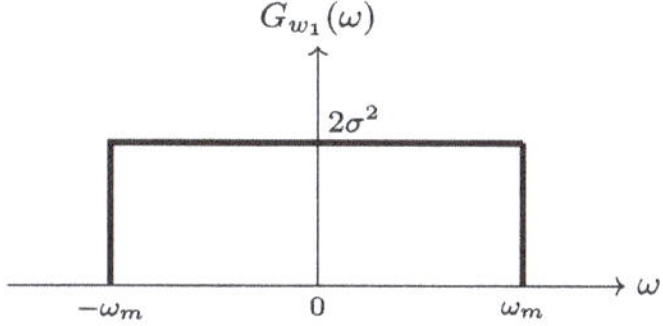

Figure 4.17: Representation of $G_{w_1}(\omega)$

The discriminator output in Fig. 4.16 is given by

$$g(t) = \begin{cases} \psi_0(t) = k_p f(t) + \frac{1}{E_c} w_1(t) & , \quad PM \\ \psi_0'(t) = k_F f(t) + \frac{1}{E_c} w_1'(t) & , \quad FM \end{cases} \tag{4.124}$$

Noise Analysis in PM and FM

$$w_1(t) \longrightarrow \boxed{\frac{d(\cdot)}{dt} \leftrightarrow j\omega} \longrightarrow w_1'(t)$$

$$G_1(\omega) = |j\omega|^2 G_{w_1}(\omega) = 2\sigma^2\omega^2 \quad , \quad |\omega| < \omega_n \tag{4.125}$$

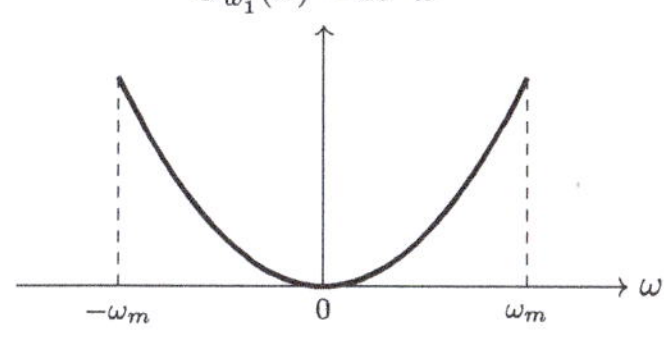

Figure 4.18: $G_{w_2}(\omega)$

Post-filter in Fig. 4.17 is used to further limit the noise in Fig. 4.16. From (4.124), post-filter allows the message to pass through without distortion, while limiting the noise bandwidth to $(-B_0, B_0)$.

Noise Analysis in PM and FM

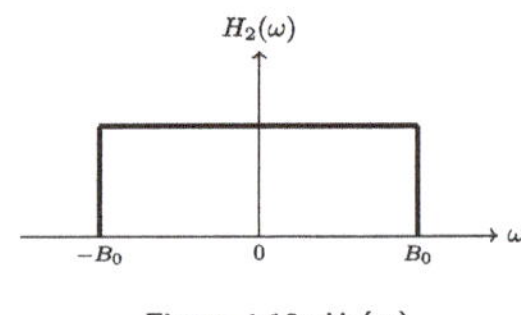

Figure 4.19: $H_2(\omega)$

Hence from (4.124), post-filter output has the form

$$g_0(t) = \begin{cases} k_p f(t) + \frac{1}{E_c} w_2(t) & , \quad PM \\ k_F f(t) + \frac{1}{E_c} w_2'(t) & , \quad FM \end{cases} \tag{4.126}$$

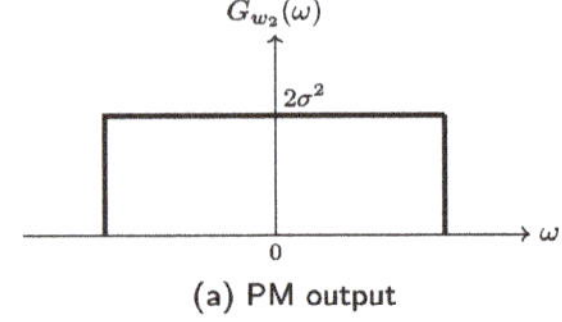

(a) PM output

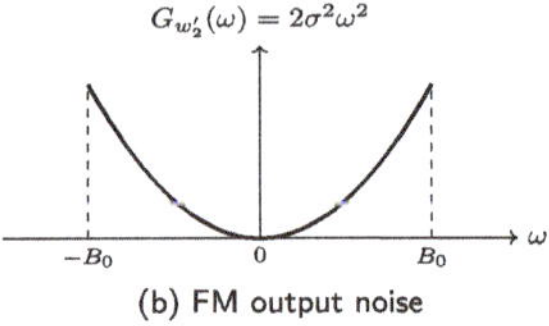

(b) FM output noise

Figure 4.20: Post-filter output

Noise Analysis in PM and FM

PM

$$\begin{aligned} SNR_o &= \frac{{k_p}^2\overline{f^2(t)}}{E[{w_2}^2(t)/{E_c}^2]} = \frac{{E_c}^2{k_p}^2\overline{f^2(t)}}{\int_{-B_0}^{B_0} G_{w_2}(\omega)d\omega} \\ &= \frac{{k_p}^2\overline{f^2(t)}}{2\sigma^2 \times 2B_0/{E_c}^2} = \frac{{E_c}^2{k_p}^2\overline{f^2(t)}}{4\sigma^2 B_0} \end{aligned} \tag{4.127}$$

For a single-tone signal

$$f(t) = E_0 \sin B_0 t \tag{4.128}$$

$$\overline{f^2(t)} = {E_0}^2/2 \tag{4.129}$$

$$SNR_o = \frac{{E_c}^2(E_0 k_p)^2}{8\sigma^2 B_0} = \frac{{E_c}^2\beta^2}{8\sigma^2 B_0} \tag{4.130}$$

Using (4.119), the PM signal efficiency is given by

$$\eta_{PM} = \frac{SNR_o}{SNR_i} = \frac{{E_c}^2\beta^2}{8\sigma^2 B_0} \cdot \frac{8\sigma^2\beta B_0}{{E_c}^2} = \beta^3 \tag{4.131}$$

Thus the SNR gain $\propto \beta^3$ for PM modulation scheme.

Noise Analysis in PM and FM

FM

$$SNR_o = \frac{k_F{}^2\overline{f^2(t)}}{E\left(w_2'(t)\right)^2/E_c{}^2} = \frac{E_c{}^2 k_F{}^2\overline{f^2(t)}}{E\left(w_2'(t)\right)^2} \tag{4.132}$$

$$E\left[(w_2'(t))^2\right] = \int_{-B_0}^{B_0} G_{w_2'}(\omega)\, d\omega = 2\sigma^2 \int_{-B_0}^{B_0} \omega^2\, d\omega = \frac{4\sigma^2 B_0{}^3}{3} \tag{4.133}$$

Using (4.132) in (4.133)

$$SNR_o = \frac{3E_c{}^2 k_F{}^2\overline{f^2(t)}}{4\sigma^2 B_0{}^3} \tag{4.134}$$

For a single tone as in (4.128)

$$SNR_o = \frac{3E_c{}^2(k_F E_0)^2}{8\sigma^2 B_0{}^3} = \frac{3E_c{}^2}{8\sigma^2 B_0}\left(\frac{k_F E_0}{B_0}\right)^2 = \frac{3E_c{}^2\beta^2}{8\sigma^2 B_0} \tag{4.135}$$

Thus the SNR gain for FM is given by

$$\eta_{FM} = \frac{SNR_o}{SNR_i} = \frac{3E_c{}^2\beta^2}{8\sigma^2 B_0}\cdot\frac{8\sigma^2\beta B_0}{E_c{}^2} = 3\beta^3 \tag{4.136}$$

Fourier Series Fourier Transforms Amplitude Modulation Angle Modulation Digital Modulation Techniques

Noise Analysis in PM and FM

Thus

$$\eta = \frac{SNR_o}{SNR_i} = \begin{cases} \beta^3 & , \quad PM \\ 3\beta^3 & , \quad FM \end{cases} \tag{4.137}$$

Figure 4.21: SNR gain of PM and FM modulation schemes

For PM/FM, SNR gain $\propto \beta^3$, whereas for AM it is a factor of 2.
Since

$$\beta = \begin{cases} k_p E_0 & , \quad PM \\ \frac{k_F E_0}{B_0} & , \quad FM \end{cases} \tag{4.138}$$

for PM/FM, improvement in SNR can be realized by adjusting the above transmit parameters. This is unlike AM, where the gain is constant at $3dB$.

5 Digital Modulation Techniques

Abstract

This chapter provides an introduction to Digital Modulation Techniques covering topics such as sampling, quantization (Lloyd-Max and Uniform), elements of coding and modulation theory which are the building blocks of modern digital communication systems. Finally the classic problem of detection of signal in noise is analyzed in the Binary Pulse Code Modulation (PCM) signaling case, using the minimization of the probability of error criterion. As a follow up, the optimum waveform selection in the PCM case is completely solved here. This is followed by an analysis of the M-ary orthogonal signaling and the corresponding probability of error calculations. PCM analysis shows that the probability of error can be made arbitrarily small through a combination of coding/modulation schemes, and as a result error-free transmission is possible to any desired degree of accuracy.

U. Pillai and A. Patel, *Fundamentals of Communication Theory*,
https://doi.org/10.1007/978-3-032-20615-2_5

Table of Contents

1. Fourier Series

2. Fourier Transforms

3. Amplitude Modulation

4. Angle Modulation

5. Digital Modulation Techniques
5.1 Digital Modulation Techniques
5.2 Lloyd Max Quantizer
5.3 Matched Filter Receiver
5.4 Matched Filter as Correlator Receiver
5.5 Pulse Code Modulation (PCM) Noise Analysis
5.6 Waveform Design and Modulation Schemes
5.7 Probability of Error Analysis for M-ary Orthogonal Signaling Scheme

Digital Modulation Techniques

- Bandlimited signals $f(t)$ can be sampled to generate $f(kT)$ without loosing information.
- Sampled signal values $f(kT)$ can be quantized to $\hat{f}(kT)$ that belongs to a finite alphabet set $\{y_1, y_2, \ldots, y_n\}$ to achieve any degree of accuracy.
- The quantized output $\hat{f}(kT)$ correspond to a finite alphabet set $\{y_1, y_2, \ldots, y_n\}$ with n symbols and can be binary represented with m bits, provided $n \leq 2^m$.
- The binary output stream can be boosted through coding to increase redundancy or robustness for error correction.
- Each binary bit has two states (1 or 0) and can be transmitted using waveforms $s_1(t)$ and $s_0(t)$ or other combinations by grouping several bits together. These ideas are summarized below in Fig. 5.1

Digital Modulation Techniques

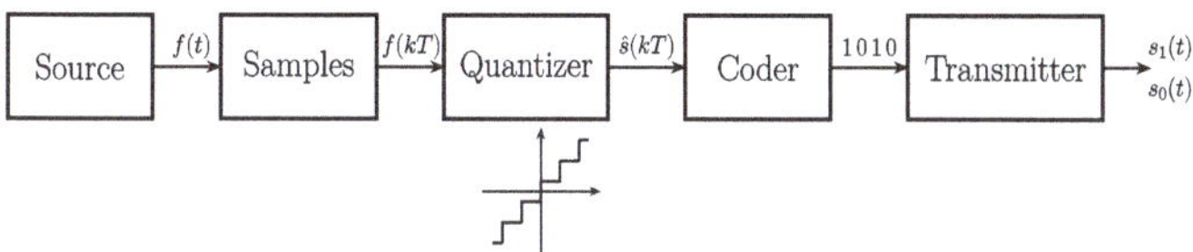

Figure 5.1: Digital Transmission Scheme

From Nyquist's Sampling Theorem, if $f(t)$ is bandlimited to $(-B_0, B_0)$ i.e.,

$$F(\omega) = 0 \quad , \quad |\omega| > B_0,$$

then $f(t)$ can be reconstructed from its samples $f(kT)$ as

$$f(t) = \sum f(kT)\,\text{sinc}(\omega_0(t - kT)) \quad , \quad \omega_0 = \frac{2\pi}{T} \tag{5.1}$$

provided the sampling interval T satisfies

$$T \leq T_N = \frac{\pi}{B_0} = \frac{1}{2f_s} \tag{5.2}$$

Fourier Series Fourier Transforms Amplitude Modulation Angle Modulation Digital Modulation Techniques

Digital Modulation Techniques

From (5.1), the set $\{f(kT)\}_{k=-\infty}^{\infty}$ is equivalent to the continuous signal $f(t)$ and can be transmitted in place of $f(t)$.

The sample values $f(kT)$ in reality range over a continuum of values such as $-M < f(kT) < M$.

They can be discretized to a finite set of values $\{y_1, y_2, \ldots, y_n\}$ by partitioning the signal range $(-M, M)$ into n segments by segmenting at $x_1, x_2, \ldots, x_{n-1}$ as shown in Fig. 5.2

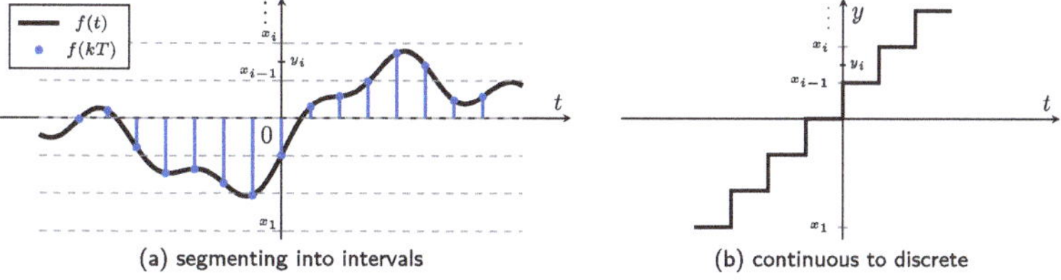

(a) segmenting into intervals (b) continuous to discrete

Figure 5.2: Quantization

Digital Modulation Techniques

Generally, all signal values $f(kT)$ are not of equal importance, and this can be captured through its histogram $f_x(x)$ shown in Fig. 5.3

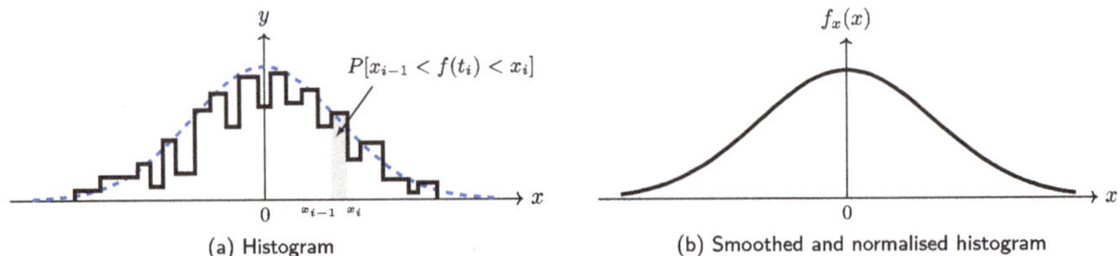

(a) Histogram (b) Smoothed and normalised histogram

Figure 5.3: Signal Histogram

In Fig. 4.1 b, $f_x(x)$ represents the smoothed and normalized histogram of the underlying signal $f(t)$. Thus

$$\int_{-\infty}^{+\infty} f_x(x)\, dx = 1 \tag{5.3}$$

and $\int_{x_{i-1}}^{x_i} f_x(x)\, dx$ represents the likelihood that the sampled signal $f(kT)$ will fall in the region (x_{i-1}, x_i).

Digital Modulation Techniques

Hence, the quantization problem can be stated as follows:
Determine $\{x_i\}_{i=1}^{n-1}$ and $\{y_i\}_{i=1}^{n}$ in some optimal manner such that, if

$$x_{i-1} < f(kT) < x_i \Rightarrow x = \hat{f}(kT) = y_i \tag{5.4}$$

Thus the input $x = f(kT)$ is assigned the value $y = \hat{f}(kT)$ as shown in (5.4).

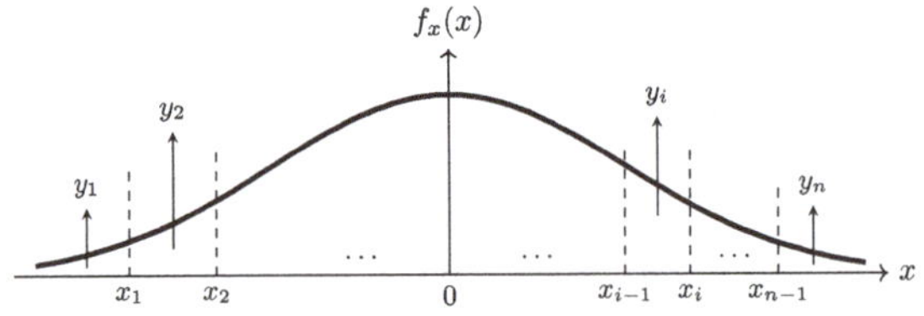

Figure 5.4: Optimal Quantization

Lloyd Max Quantizer

If we select the overall goal as minimizing the mean square error D, then using Fig. 5.4 , this gives

$$D = E\{(x-y)^2\} = \sum_{i=1}^{n} \int_{x_{i-1}}^{x_i} (x-y_i)^2 f_x(x)\,dx$$
$$= \int_{-\infty}^{x_1} (x-y_1)^2 f_x(x)\,dx + \ldots + \int_{x_{i-1}}^{x_i} (x-y_i)^2 f_x(x)\,dx + \quad (5.5)$$
$$\int_{x_i}^{x_{i+1}} (x-y_{i+1})^2 f_x(x)\,dx + \ldots + \int_{x_{n-1}}^{\infty} (x-y_n)^2 f_x(x)\,dx$$

To minimize D, differentiate D in (5.5) with respect to x_i and y_i respectively. This gives

$$\frac{\partial D}{\partial x_i} = (x_i - y_i)^2 f_x(x_i) - (x_i - y_{i+1})^2 f_x(x_i) = 0 \quad (5.6)$$

or

$$x_i{}^2 + y_i{}^2 - 2x_i y_i - (x_i{}^2 + y_{i+1}^2 - 2x_i y_{i+1}) = 0 \quad (5.7)$$

Fourier Series Fourier Transforms Amplitude Modulation Angle Modulation Digital Modulation Techniques

Lloyd Max Quantizer

which gives

$$x_i = \frac{y_i^2 - y_{i+1}^2}{2(y_i - y_{i+1})} = \frac{y_i + y_{i+1}}{2} \quad (5.8)$$

Similarly from (5.5)

$$\frac{\partial D}{\partial y_i} = \int_{x_{i-1}}^{x_i} (-2)(x-y_i) f_x(x)\,dx = 0 \quad (5.9)$$

or

$$y_i = \frac{\int_{x_{i-1}}^{x_i} x\, f_x(x)\,dx}{\int_{x_{i-1}}^{x_i} f_x(x)\,dx} = E\left[x \middle| x_{i-1} < x < x_i\right] \quad (5.10)$$

Equations (5.8) and (5.10) represent a coupled set of non-linear equations (Lloyd-Max Equations) that can be alternately iterated to obtain the optimum quantization values x_i and quantization levels y_i. Convergence of this alternate-iterative procedure is guaranteed from the convexity properties of the individual solutions in (5.8) and (5.10).

Lloyd Max Quantizer

In (5.8), the partition value x_i is the mean of the outputs of y_i and y_{i+1}, where as in (5.10), the output y_i is the conditional mean of x given that $x_{i-1} < x < x_i$.

Special Case: Uniform Output Quantizer
If we assume

$$P[\hat{f}(kT) = y_i] = \frac{1}{n} \quad , \quad i = 1 \to n \tag{5.11}$$

then

$$P(Y = y_i) = P[x_{i-1} < x < x_i] = \int_{x_{i-1}}^{x_i} f_x(x)\, dx = \frac{1}{n} \tag{5.12}$$

In that case, (5.12) can be used to determine the boundary values of x_i, and knowing x_i, (5.10) can be used to determine the y_i's.
We illustrate this through and example.

Lloyd Max Quantizer

Example: Design a four-level uniform output quantizer for a triangular histogram in $(-1, 1)$.

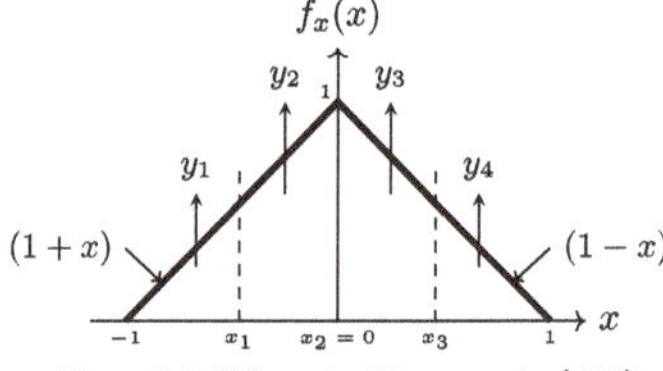

Figure 5.5: Triangular Histogram in (-1,1)

From the symmetry of $f_x(x)$, we have $x_1 = -x_3$, $x_2 = 0$, $y_2 = -y_3$, $y_1 = -y_4$. To obtain x_3, using (5.12)

$$\int_0^{x_3} f_x(x)\, dx = \int_0^{x_3} (1 - x)\, dx = \left(x - \frac{x^2}{2}\right)\Bigg|_0^{x_3} = \frac{1}{4}$$

or

$$2{x_3}^2 - 4x_3 + 1 = 0 \Rightarrow x_3 = \frac{4 \pm \sqrt{16 - 8}}{4} = 1 - \frac{1}{\sqrt{2}} \approx 0.2929 \tag{5.13}$$

Lloyd Max Quantizer

From (5.10)

$$y_3 = \frac{\int_0^{x_3} x\, f_x(x)\, dx}{1/4} = 4\int_0^{x_3} x(1-x)\, dx = 4\left(\frac{x^2}{2} - \frac{x^3}{3}\right)\Bigg|_0^{0.2929} \approx 0.1381 \quad (5.14)$$

and

$$y_4 = \frac{\int_{x_3}^{x_4} x(1-x)\, dx}{1/4} = 4\left(\frac{x^2}{2} - \frac{x^3}{3}\right)\Bigg|_{0.2929}^{1} \approx 0.5286 \quad (5.15)$$

Table 5.1: x_i and y_i values

x_i	y_i
$x_1 = -0.2929$	$y_1 = -0.5286$
$x_2 = 0$	$y_2 = -0.1381$
$x_2 = 0.2929$	$y_3 = 0.1381$
	$y_4 = 0.5286$

Lloyd Max Quantizer

(a) Original Image

(b) 3-bit Lloyd Max Quantizer

(c) 4-bit Lloyd Max Quantizer

Figure 5.6: Lloyd Max Quantization on "Peppers" image. Image credits: Sai Meghana Kuchana - used with permission.

Table 5.2: Lloyd–Max Error

Image	**N = 3 bit (8-level)**	**N = 4 bit (16-level)**
House	24.837	7.605
Mandrill	42.108	11.186
Peppers	54.025	13.667
Female	26.603	6.627

Coder

The quantizer output $\hat{f}(kT)$ in Fig. 5.1 is assigned to a finite alphabet set $\{y_1, y_2, \ldots, y_n\}$ and the simplest way to code using a binary alphabet is to assign a codeword of m bits consisting of 1's and 0's to each y_i using some rule as follows:

$$y_k = \underset{\substack{\uparrow \\ 1}}{1}0110\ldots\underset{\substack{\uparrow \\ m}}{1} \quad , \quad k = 1 \to n \tag{5.16}$$

and this is always possible if $n \leq 2^m$. If $N = 2^m$ is much larger than n, then this gives the option of assigning each y_i to a codeword such that there is reasonable separation between the assigned words.

Coder

For example, if the output alphabet is only $\{y_1, y_2\}$, then we may assign

$$y_1 \to 0 \quad , \quad y_2 \to 1 \quad \text{using one bit}$$

or

$$y_1 \to 00 \quad , \quad y_2 \to 11 \quad \text{using two bits} \tag{5.17}$$

or

$$y_1 \to 000 \quad , \quad y_2 \to 111 \quad \text{using three bits}$$

The second and third possibilities above are opted to address error correcting at the receiver output, since in the later case any single error at any one location such as $100, 001$ or 001 can be easily identified and corrected as 000.
For our discussion, we shall assume the coder output in Fig. 5.1 is a binary string of 1s and 0s.
Finally the transmitter assigns waveform to the string of 1's and 0's at the coder output.

Transmitter

Once again, the simplest way to assign waveforms to binary data is to assign

$$\begin{aligned} 1 &\rightarrow s_1(t) \\ 0 &\rightarrow s_0(t) \end{aligned} \tag{5.18}$$

or more generally, two bits can be grouped together and assigned waveforms as follows

$$\begin{aligned} 0\,0 &\rightarrow s_0(t) \\ 0\,1 &\rightarrow s_1(t) \\ 1\,0 &\rightarrow s_2(t) \\ 1\,1 &\rightarrow s_3(t) \end{aligned} \tag{5.19}$$

Obviously this procedure can be generalized to 4 bits and 16 waveforms or 8 bits and their associated 256 waveforms.
The problem of optimally identifying the waveforms in each such cases under some restrictions such as total transmit energy, pulse length etc. come under waveform design.

Transmitter

In the simplest case, the waveforms design problem is to select $s_1(t)$ and $s_0(t)$ in (5.18) such that the overall probability of error P_ϵ is minimized. Waveform design is relevant only because transmitted signal is affected by noise as shown in Fig. 5.7 and towards the goal of error-free detection, the next question is to address the receiver design problem for a known signal $s(t)$ that is affected by noise $n(t)$.

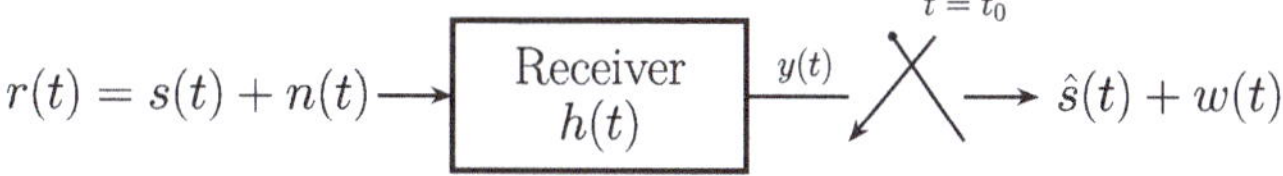

Figure 5.7: "Best Receiver" in signal plus noise

Matched Filter Receiver

A known signal $s(t)$ of energy E that is contaminated by uncorrelated white noise $n(t)$ is applied to a receiver with unknown impulse response $h(t)$ as shown in Fig. 5.8

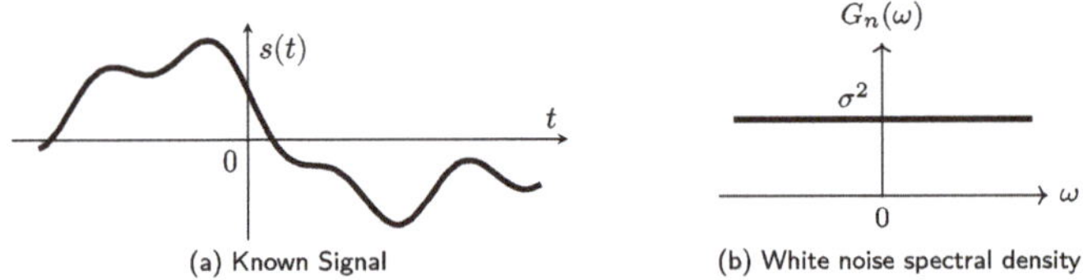

Figure 5.8: Known signal corrupted with uncorrelated white noise

This gives the signal output to be

$$\hat{s}(t) = s(t) \circledast h(t) \leftrightarrow S(\omega)\, H(\omega) \tag{5.20}$$

and the output noise to be

$$w(t) = n(t) \circledast h(t) = \int_{-\infty}^{\infty} n(\tau)h(t-\tau)\, d\tau \tag{5.21}$$

Matched Filter Receiver

To obtain the "best receiver filter" $h(t)$, we can maximize the output signal-to-noise ratio (SNR_o). Define

$$SNR_o = \frac{\text{Signal power at } t = t_0}{\text{Average output noise power}} = \frac{|\hat{s}(t_0)|^2}{E\left[w^2(t)\right]} \tag{5.22}$$

From (5.20),

$$s(t) = \frac{1}{2\pi}\int_{-\infty}^{\infty} S(\omega)H(\omega)e^{j\omega t}\, d\omega \tag{5.23}$$

so that

$$s(t_0) = \frac{1}{2\pi}\int_{-\infty}^{\infty} S(\omega)H(\omega)e^{j\omega t_0}\, d\omega \tag{5.24}$$

Matched Filter Receiver

and from (5.21), the average output noise power is given by

$$\begin{aligned} E\left[w^2(t)\right] &= E\left[\left|\int_{-\infty}^{\infty} n(\tau)h(t-\tau)\,d\tau\right|^2\right] \\ &= E\left[\int_{-\infty}^{\infty} n(\tau_1)h(t-\tau_1)\,d\tau_1 \cdot \int_{-\infty}^{\infty} n(\tau_2)h(t-\tau_2)\,d\tau_2\right] \\ &= \int_{-\infty}^{\infty}\int_{-\infty}^{\infty} E\left[n(\tau_1)n(\tau_2)\right] h(t-\tau_1)h(t-\tau_2)\,d\tau_1\,d\tau_2 \end{aligned} \tag{5.25}$$

Since $n(t)$ is uncorrelated white noise

$$E\left[n(\tau_1)n(\tau_2)\right] = \sigma^2\delta(\tau_1-\tau_2) \tag{5.26}$$

Matched Filter Receiver

With (5.26) in (5.25),

$$\begin{aligned} E[w^2(t)] &= \sigma^2 \int_{-\infty}^{\infty}\int_{-\infty}^{\infty} \delta(\tau_1-\tau_2)\,h(t-\tau_1)\,h(t-\tau_2) \\ &= \sigma^2 \int_{-\infty}^{\infty} |h(t-\tau)|^2\,d\tau = \sigma^2 \int_{-\infty}^{\infty} |h(t)|^2\,dt \\ &= \frac{\sigma^2}{2\pi}\int_{-\infty}^{\infty} |H(\omega)|^2\,d\omega \end{aligned} \tag{5.27}$$

where we have used Parseval's theorem. Substituting (5.24) and (5.27) into (5.22), we get

$$\begin{aligned} SNR_o &= \frac{\left|\frac{1}{2\pi}\int_{-\infty}^{\infty} S(\omega)H(\omega)e^{j\omega t_0}\,d\omega\right|^2}{\frac{\sigma^2}{2\pi}\int_{-\infty}^{\infty}|H(\omega)|^2\,d\omega} \\ &= \frac{1}{2\pi\sigma^2}\frac{\left|\int_{-\infty}^{\infty} S(\omega)H(\omega)e^{j\omega t_0}\,d\omega\right|^2}{\int_{-\infty}^{\infty}|H(\omega)|^2\,d\omega} \end{aligned} \tag{5.28}$$

Matched Filter Receiver

The goal is to maximize SNR_o over the unknown $h(t) \leftrightarrow H(\omega)$. Towards this, we can make use of Schwarz' inequality, which gives

$$\left| \int_{-\infty}^{\infty} A(\omega) B(\omega)\, d\omega \right|^2 \leq \int_{-\infty}^{\infty} |A(\omega)|^2\, d\omega \cdot \int_{-\infty}^{\infty} |B(\omega)|^2\, d\omega \tag{5.29}$$

with equality in (5.29) if and only if

$$B(\omega) = A^*(\omega) \tag{5.30}$$

With $A(\omega) = S(\omega)e^{j\omega t_0}$, $B(\omega) = H(\omega)$ in (5.28); applying (5.29) we get

$$\begin{aligned} SNR_o &\leq \frac{\int_{-\infty}^{\infty} \left|S(\omega)e^{j\omega t_0}\right|^2 d\omega \int_{-\infty}^{\infty} \left|H(\omega)e^{j\omega t_0}\right|^2 d\omega}{2\pi\sigma^2 \int_{-\infty}^{\infty} |H(\omega)|^2\, d\omega} \\ &= \frac{1}{\sigma^2}\frac{1}{2\pi}\int_{-\infty}^{\infty} |S(\omega)|^2\, d\omega = \frac{1}{\sigma^2}\int_0^T s^2(t)\, dt = \frac{E}{\sigma^2} = SNR_{o,\,max} \end{aligned} \tag{5.31}$$

Matched Filter Receiver

From (5.31), clearly the maximum value of SNR is given by E/σ^2 and from (5.29) it can be achieved iff

$$H(\omega) = \left(S(\omega)e^{j\omega t_0}\right)^* = S^*(\omega)e^{-j\omega t_0} \tag{5.32}$$

or

$$h(t) = s^*(t_0 - t) \tag{5.33}$$

In summary, (5.33) represents the optimum receiver that maximizes the output SNR in (5.28) at $t_0 = t$. Since the receiver is matched to the input waveform $s(t)$, it is known as Matched Filter (MF) receiver (see Fig. 5.9). Notice that the numerator in (5.22) represents the instantaneous output signal power at $t = t_0$ as opposed to averaged power, whereas the denominator represents the average output noise power.

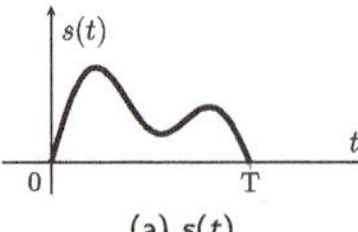

(a) $s(t)$

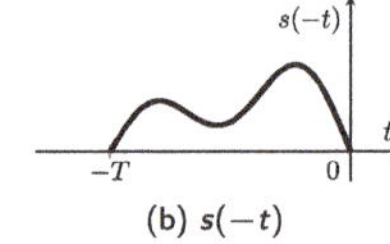

(b) $s(-t)$

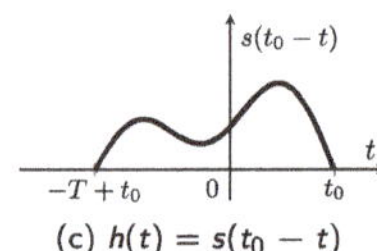

(c) $h(t) = s(t_0 - t)$

Figure 5.9: A signal and its matched filter

Matched Filter as Correlator Receiver

From Fig. 5.7, with $h(t)$ representing the optimum matched filter as in (5.33), we get the filter output to be

$$\begin{aligned} y(t) &= r(t) \circledast h(t) = r(t) \circledast s(t_0 - t) \\ &= \int_0^t r(t-\tau)s(t_0-\tau)\,d\tau \end{aligned} \tag{5.34}$$

Hence

$$y(t_0) = \int_0^{t_0} r(t_0-\tau)s(t_0-\tau)\,d\tau \tag{5.35}$$

Let $t_0 - \tau = t$, so that (5.35) reduces to

$$y(t_0) = \int_0^{t_0} r(t)\,s(t)\,dt \tag{5.36}$$

which can be implemented as in Fig. 5.10

Matched Filter as Correlator Receiver

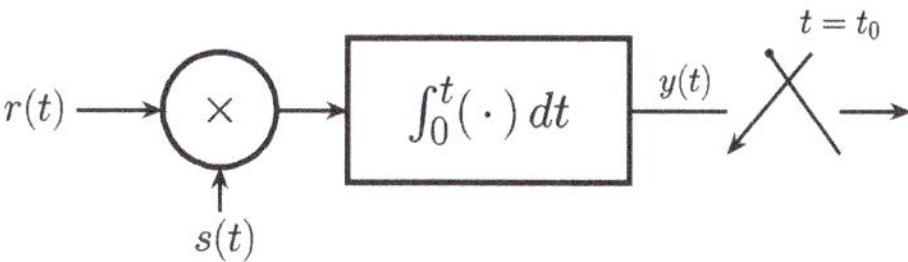

Figure 5.10: Correlator Receiver

Fig. 5.10 represents a correlator receiver where the input signal $r(t)$ is correlated with the desired signal $s(t)$ and integrated over $(0, t)$ prior to making a decision at $t = t_0$.

From the above analysis, matched filter (MF) is equivalent to a correlator receiver and hence it may be implemented as in Fig. 5.10

We can make use of the matched filter (MF) receiver/correlator to analyze the binary pulse code modulation scheme discussed in (5.18) in the presence of uncorrelated white noise.

Pulse Code Modulation (PCM) Noise Analysis

In PCM, data bit 1 at the coder output is assigned a waveform $s_1(t)$, $0 < t < T$, and bit 0 is assigned a waveform $s_0(t)$, $0 < t < T$. Thus the transmission scheme is

$$s(t) = \begin{cases} s_1(t) & , \quad H_1 \\ s_0(t) & , \quad H_0 \end{cases} \tag{5.37}$$

Here H_1 and H_0 represent the two "hypotheses" that "1 is transmitted" and "0 is transmitted" respectively. and

$$\int_0^T s_1{}^2(t)\,dt = \int_0^T s_0{}^2(t)\,dt = E \tag{5.38}$$

Figure 5.11: PCM Modulator

Fourier Series Fourier Transforms Amplitude Modulation Angle Modulation Digital Modulation Techniques

Pulse Code Modulation (PCM) Noise Analysis

The channel adds zero mean uncorrelated white Gaussian noise $n(t)$ of spectral density σ^2 so that the received signal $r(t)$ equals

$$r(t) = \begin{cases} s_1(t) + n(t) & , \quad H_1 \\ s_0(t) + n(t) & , \quad H_0 \end{cases} \tag{5.39}$$

Since the optimum receiver for H_1 is the correlator receiver in Fig. 5.10 the receiver in Fig. 5.12 is optimum in maximizing the output SNR for $s_1(t)$.

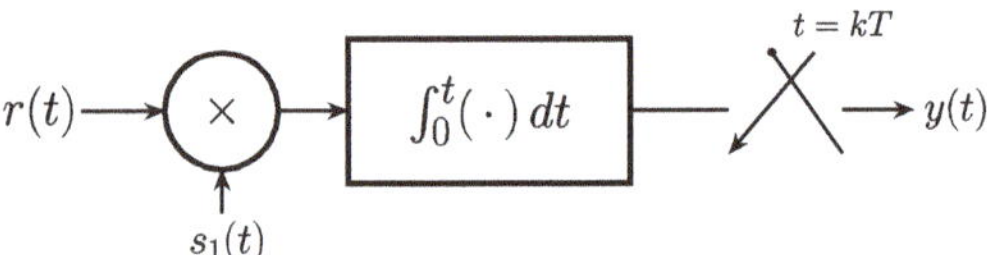

Figure 5.12: Correlator Receiver

Note that the receiver in Fig. 5.12 is matched to $s_1(t)$.

Fundamentals of Communication Theory U. Pillai & A. Patel

Pulse Code Modulation (PCM) Noise Analysis

Hence the output y at $t = T$ can be expressed as

$$y = \int_0^T r(t)s_1(t)\,dt \tag{5.40}$$

Under H_1 in (5.39), (5.40) reduces to

$$\begin{aligned} y &= \int_0^T (s_1(t) + n(t))\, s_1(t)\,dt = \int_0^T {s_1}^2(t)\,dt + \int_0^T s_1(t)\,n(t)\,dt \\ &= E + W \quad , \quad H_1 \end{aligned} \tag{5.41}$$

and under H_0 in (5.39), (5.40) reduces to

$$\begin{aligned} y &= \int_0^T (s_0(t) + n(t))\, s_1(t)\,dt = \int_0^T s_1(t)\,s_0(t)\,dt + \int_0^T s_1(t)\,n(t)\,dt \\ &= \rho E + W \quad , \quad H_0 \end{aligned} \tag{5.42}$$

Fourier Series Fourier Transforms Amplitude Modulation Angle Modulation Digital Modulation Techniques

Pulse Code Modulation (PCM) Noise Analysis

Here ρ represents the correlation coefficient between $s_1(t)$ and $s_0(t)$ defined by

$$\rho = \frac{\int_0^T s_1(t)\,s_0(t)\,dt}{\sqrt{\int_0^T {s_1}^2(t)\,dt} \cdot \sqrt{\int_0^T {s_0}^2(t)\,dt}} = \frac{\int_0^T s_1(t)\,s_0(t)\,dt}{E} \tag{5.43}$$

or

$$\int_0^T s_1(t)\,s_0(t)\,dt = \rho E \tag{5.44}$$

Using (5.41)-(5.42) in (5.40), we get

$$y = \begin{cases} E + W & , \quad H_1 \\ \rho E + W & , \quad H_0 \end{cases} \tag{5.45}$$

where

$$W = \int_0^T s_1(t)\,n(t)\,dt \tag{5.46}$$

represents the common output noise component.

Pulse Code Modulation (PCM) Noise Analysis

Fig. 5.13 shows the received data vector y in (5.45) under the two hypotheses.

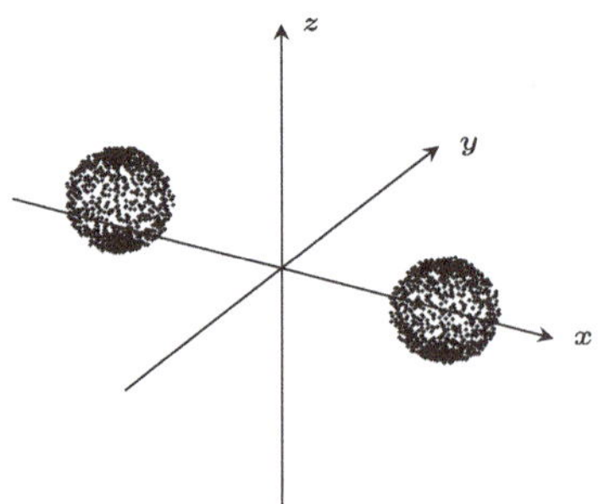

Figure 5.13: Received data vector

If $n(t)$ represents zero mean uncorrelated Gaussian noise, then

$$E[n(t)] = 0,\ E[n^2(t)] = \sigma^2,\ E[n(t_1)n(t_2)] = 0 \quad , \quad t_1 \neq t_2 \tag{5.47}$$

Fourier Series Fourier Transforms Amplitude Modulation Angle Modulation Digital Modulation Techniques

Pulse Code Modulation (PCM) Noise Analysis

Since W is linearly related to $n(t)$ in (5.46), W represents a Gaussian random variable with

$$E(W) = \int_0^T s_1(t)\,\underbrace{E[n(t)]}_{0}\,dt = 0 \tag{5.48}$$

and variance

$$\begin{aligned}
{\sigma_W}^2 = E[W^2] &= E\left[\left(\int_0^T s_1(t)n(t)\,dt\right)^2\right] \\
&= E\left[\int_0^T s_1(t_1)n(t_1)\,dt_1 \cdot \int_0^T s_1(t_2)n(t_2)\,dt_2\right] \\
&= \int_0^T \int_0^T s_1(t)s_2(t)\,\underbrace{E[n(t_1)n(t_2)]}_{\sigma^2\delta(t_1-t_2)}\,dt_1\,dt_2 \\
&= \int_0^T {s_1}^2(t)\,\underbrace{E[n^2(t)]}_{\sigma^2}\,dt = \sigma^2 \int_0^T {s_1}^2(t)\,dt = \sigma^2 E
\end{aligned} \tag{5.49}$$

Fundamentals of Communication Theory U. Pillai & A. Patel

Pulse Code Modulation (PCM) Noise Analysis

From (5.48) and (5.49)

$$W \sim \mathcal{N}(0, \sigma^2 E) \tag{5.50}$$

so that from (5.45), y is also Gaussian random variable with mean values E and σE under H_1 and H_0 respectively and with common variance $\sigma^2 E$. Hence

$$y \sim \begin{cases} \mathcal{N}(E, \sigma^2 E) & , \quad H_1 \\ \mathcal{N}(\rho E, \sigma^2 E) & , \quad H_0 \end{cases} \tag{5.51}$$

or the respective probability density functions (pdfs) are given by

$$f_{y|H_1}(y|H_1) = \frac{1}{\sqrt{2\pi\sigma^2 E}} e^{-(y-E)^2/2\sigma^2 E} \tag{5.52}$$

and

$$f_{y|H_0}(y|H_0) = \frac{1}{\sqrt{2\pi\sigma^2 E}} e^{-(y-\rho E)^2/2\sigma^2 E} \tag{5.53}$$

Fourier Series Fourier Transforms Amplitude Modulation Angle Modulation Digital Modulation Techniques

Pulse Code Modulation (PCM) Noise Analysis

These pdfs are shown in Fig. 5.14

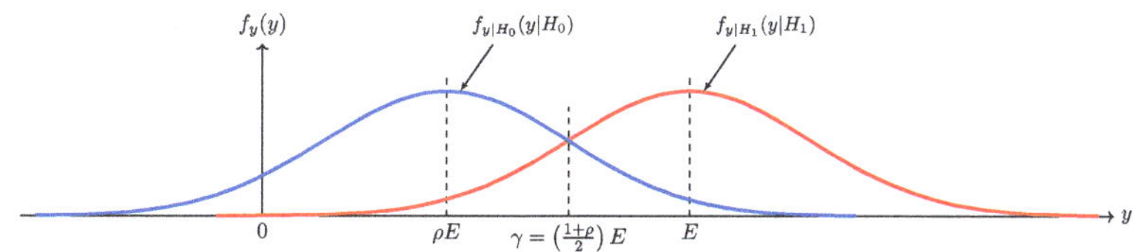

Figure 5.14: $f_{y|H_1}(y|H_1)$ and $f_{y|H_0}(y|H_0)$

$f_{y|H_1}(y|H_1)$ represents the "likelihood" of bit "1" spreading out under the noise variable W, and similarly $f_{y|H_0}(y|H_0)$ represents the "likelihood" of bit "0" spreading out under the noise W.
Their respective peak locations at E and ρE represent the "most-likely" values of the output y under "no-noise" conditions.

Fundamentals of Communication Theory U. Pillai & A. Patel

Pulse Code Modulation (PCM) Noise Analysis

Hence their midpoint η is given by (see Fig. 5.14)

$$\eta = \left(\frac{1+\rho}{2}\right) E \tag{5.54}$$

may be chosen as a reasonable (in fact, optimum) threshold for decision making. Thus we may choose the following criterion for detection

$$\begin{aligned} y > \eta &\Rightarrow \text{decoded data bit} = 1 \\ y < \eta &\Rightarrow \text{decoded data bit} = 0 \end{aligned} \tag{5.55}$$

From (5.55), when output y at $t = kT$ exceeds the threshold η, then the corresponding data $r(t)$, $(k-1)T < t \leq kT$ is estimated as the bit "1", and if $y < \eta$, the underlying data is estimated as bit "0".

Fourier Series Fourier Transforms Amplitude Modulation Angle Modulation Digital Modulation Techniques

Pulse Code Modulation (PCM) Noise Analysis

Hence the probability of detection P_D is given by

$$\begin{aligned} P_D &= P\Bigg[\Big\{\underbrace{\text{Detect in favour of } H_1}_{H_1} \text{ and } \underbrace{H_1 \text{ is true}}_{H_1}\Big\} \text{ or } \\ &\qquad \Big\{\underbrace{\text{Detect in favour of } H_0}_{H_0} \text{ and } \underbrace{H_0 \text{ is true}}_{H_0}\Big\}\Bigg] \\ &= P[(H_1 \cap H_1) \cup (H_0 \cap H_0)] = P(H_1 \cap H_1) + P(H_0 \cap H_0) \\ &= P(H_1|H_1)\, P(H_1) + P(H_0|H_0)\, P(H_0) \end{aligned} \tag{5.56}$$

where we added the probability of mutually exclusive events. Similarly the probabilty of error P_ϵ is given by

Fundamentals of Communication Theory U. Pillai & A. Patel

Pulse Code Modulation (PCM) Noise Analysis

$$P_\epsilon = P\Big[\Big\{\underbrace{\text{Detect in favour of } H_1}_{H_1} \text{ and } \underbrace{H_0 \text{ is true}}_{H_0}\Big\} \text{ or }$$
$$\Big\{\underbrace{\text{Detect in favour of } H_0}_{H_0} \text{ and } \underbrace{H_1 \text{ is true}}_{H_1}\Big\}\Big] \tag{5.57}$$
$$= P[(H_1 \cap H_0) \cup (H_0 \cap H_1)] = P(H_1 \cap H_0) + P(H_0 \cap H_1)$$
$$= P(H_1|H_0)\,P(H_0) + P(H_0|H_1)\,P(H_1)$$
$$= P_F\,P(H_0) + P_M\,P(H_1)$$

Here

$$P_F = P(H_1|H_0) \tag{5.58}$$

represents the probability of false alarm where the data is detected as "1" when it actually represents "0". In radar, this amounts to detecting a target (H_1) when no target is present (H_0) leading to false alarm. Similarly

$$P_M = P(H_0|H_1) \tag{5.59}$$

Fourier Series Fourier Transforms Amplitude Modulation Angle Modulation Digital Modulation Techniques

Pulse Code Modulation (PCM) Noise Analysis

refers to the probability of miss, where the data is detected as "0" when it represents "1". In radar, this corresponds to declaring that no target is present (H_0) when infact the target is present (H_1) in the data, leading to missing the target. From (5.53)-(5.55) and the above discussion

$$P_F = P(H_1|H_0) = P(y > \eta \,|\, H_0) = \int_\eta^\infty f_{y|H_0}(y|H_0)\,dy \tag{5.60}$$

and

$$P_M = P(H_0|H_1) = P(y < \eta \,|\, H_1) = \int_{-\infty}^\eta f_{y|H_1}(y|H_1)\,dy \tag{5.61}$$

Substituting (5.53) into (5.60), we get

$$P_F = \int_\eta^\infty f_{y|H_0}(y|H_0)\,dy = \int_\eta^\infty \frac{1}{\sqrt{2\pi\sigma^2 E}} e^{-(y-\rho E)^2/2\sigma^2 E}\,dy \tag{5.62}$$

Fundamentals of Communication Theory U. Pillai & A. Patel

Pulse Code Modulation (PCM) Noise Analysis

Let

$$\frac{y-\rho E}{\sigma\sqrt{E}} = x, \qquad \text{so that} \qquad dy = \sigma\sqrt{E}\,dx \tag{5.63}$$

and the lower limit η given by (5.54) becomes

$$\frac{\eta-\rho E}{\sigma\sqrt{E}} = \frac{\left(\frac{1+\rho}{2}\right)-\rho E}{\sigma\sqrt{E}} = \left(\frac{1-\rho}{2}\right)\sqrt{\frac{E}{\sigma^2}} = \left(\frac{1-\rho}{2}\right)\sqrt{SNR} \tag{5.64}$$

since $\frac{E}{\sigma^2} = \frac{\int_0^T {s_1}^2(t)\,dt}{\sigma^2}$ represents the input signal-to-noise ratio.

With (5.63)-(5.64) in (5.62), we get

$$P_F = \int_{\left(\frac{1-\rho}{2}\right)\sqrt{SNR}}^{\infty} \frac{1}{\sqrt{2\pi}} e^{-x^2/2}\,dx = Q\left(\left(\frac{1-\rho}{2}\right)\sqrt{SNR}\right) \tag{5.65}$$

where

$$Q(z) = \int_z^{\infty} \frac{1}{\sqrt{2\pi}} e^{-x^2/2}\,dx \tag{5.66}$$

Fourier Series Fourier Transforms Amplitude Modulation Angle Modulation Digital Modulation Techniques

Pulse Code Modulation (PCM) Noise Analysis

represents the standard error function that represents the tail area beyond z under the standard normal distribution given in Fig. 5.15

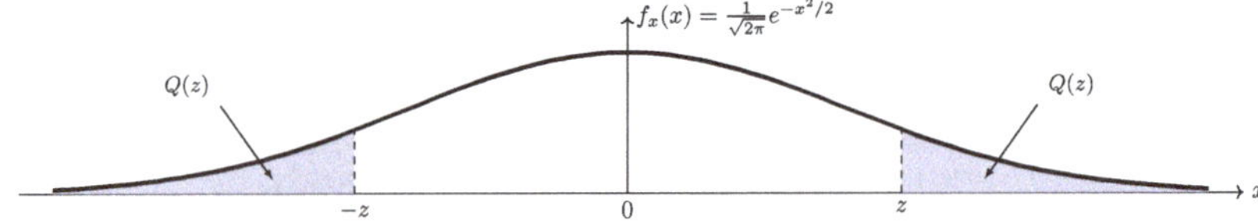

Figure 5.15: Standard Normal Distribution and error function

Similarly using (5.52) in (5.61), we get

$$P_M = \int_{-\infty}^{\eta} f_{y|H_1}(y|H_1)\,dy = \int_{-\infty}^{\eta} \frac{1}{\sqrt{2\pi\sigma^2 E}} e^{-(y-E)^2/2\sigma^2 E}\,dy \tag{5.67}$$

Let

$$\frac{y-E}{\sigma\sqrt{E}} = x, \qquad \text{so that} \qquad dy = \sigma\sqrt{E}\,dx$$

and the upper limit η given by (5.54) becomes

Fundamentals of Communication Theory U. Pillai & A. Patel

Pulse Code Modulation (PCM) Noise Analysis

$$\frac{\eta - E}{\sigma\sqrt{E}} = \frac{\left(\frac{1+\rho}{2}\right) - E}{\sigma\sqrt{E}} = -\left(\frac{1-\rho}{2}\right)\sqrt{\frac{E}{\sigma^2}} = -\left(\frac{1-\rho}{2}\right)\sqrt{SNR} \quad (5.68)$$

so that

$$\begin{aligned} P_M &= \int_{-\infty}^{-\left(\frac{1-\rho}{2}\right)\sqrt{SNR}} \frac{1}{\sqrt{2\pi}} e^{-x^2/2}\, dx = \frac{(1-\rho)\sqrt{SNR}}{2} \cdot \frac{1}{\sqrt{2\pi}} e^{-x^2/2}\, dx \\ &= Q\left(\left(\frac{1-\rho}{2}\right)\sqrt{SNR}\right) \end{aligned} \quad (5.69)$$

Finally substituting (5.65) and (5.69) into (5.68), we get the probability of error to be

$$P_\epsilon = [P(H_0) + P(H_1)]\, Q\left(\left(\frac{1-\rho}{2}\right)\sqrt{SNR}\right) = Q\left(\left(\frac{1-\rho}{2}\right)\sqrt{SNR}\right) \quad (5.70)$$

where ρ represents the signal correlation in (5.43)-(5.44) and $Q(\cdot)$ as in (5.66). Eq(5.70) represents the bit error probability in binary transmission.

Pulse Code Modulation (PCM) Noise Analysis

Fig. 5.16 shows P_ϵ in (5.70) as a function of ρ and SNR.

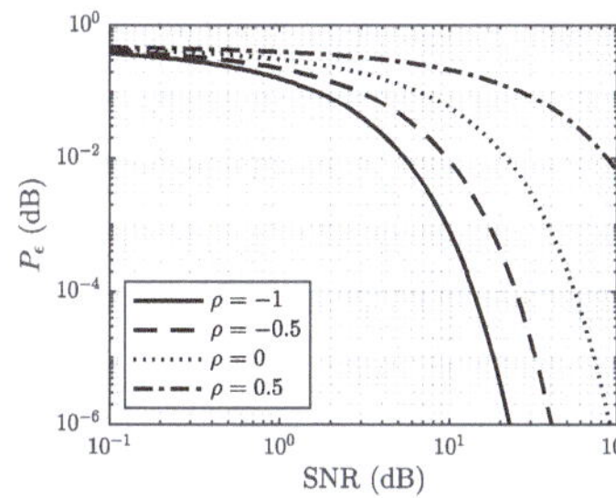

Figure 5.16: Probability of error as a function of signal correlation and SNR

From (5.70), to minimize P_ϵ, the argument $\left(\frac{1-\rho}{2}\right)\sqrt{SNR}$ must be as large as possible and this can be achieved by

1. Select ρ to be the least negative value, or

$$\Rightarrow \rho = -1 \quad (5.71)$$

2. Increase SNR by increasing the transmit signal energy E in (5.38)

Pulse Code Modulation (PCM) Noise Analysis

From (i) above, for any signal $s(t)$, $0 < t < T$, the signal pair $s_1(t) = s(t)$ and $s_0(t) = -s(t)$ is optimum in a binary signal transmission scheme to minimize the probability of error in presence of white noise, i.e., the signal shape of $s(t)$ doesn't matter in P_ϵ minimization and hence any convinient pair such as $(\cos\omega_0 t, -\cos\omega_0 t)$ can be used in binary signal transmission. The pair $(\cos\omega_0 t, -\cos\omega_0 t)$, $0 < t < T$, represents Binary Phase Shift Keying (BPSK) and is optimum in minimizing probability of error. In contrast, the orthogonal signal pair $(\cos\omega_0 t, \sin\omega_0 t)$, $0 < t < T$ is suboptimal, since ρ in that case is given by

$$\rho = \int_0^T \cos\omega_0 t \sin\omega_0 t \, dt = \frac{1}{2}\int_0^T \sin 2\omega_0 t \, dt = -\frac{1}{2}\cos 2\omega_0 t \Big|_0^T = 0$$

which leads to a higher probability of error compared to BPSK in (5.70).

Fourier Series Fourier Transforms Amplitude Modulation Angle Modulation Digital Modulation Techniques

Pulse Code Modulation (PCM) Noise Analysis

Similarly the Frequency Shift Keying Pair (FSK) pair $(\sin\omega_0 t, \sin\omega_1 t)$, $0 < t < T$ also leads to $\rho = 0$ and hence represent a suboptimal pair from the perspective of minimum probability of error transmission. These signal schemes are shown below in Fig. 5.17

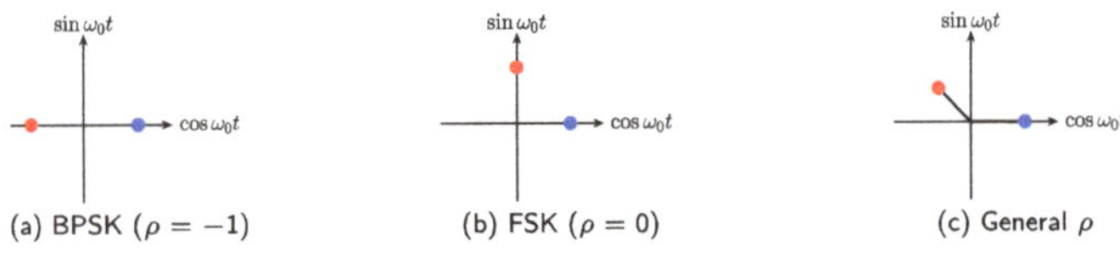

(a) BPSK ($\rho = -1$) (b) FSK ($\rho = 0$) (c) General ρ

Figure 5.17: Binary Signal transmission schemes

From (5.12), the correlation coefficient is a direct measure of "the distance" between the pair $s_1(t)$ and $s_0(t)$, and from (5.70) for minimum probability of error, the distance separation must be maximum, which appeals as reasonable from a common sense point of view.

From (ii) in (5.71), the bit error probability can also be minimized by increasing the $SNR = E/\sigma^2$, or the transmit signal energy E in (5.38).

Pulse Code Modulation (PCM) Noise Analysis

Since E has no upper bound, atleast in principle, the probability of error can be decreased to any desired level of accuracy (practically to zero) by increasing the transmit signal energy E and hence the effect of noise can be practically eliminated.

This is unlike analog transmission, where by increasing transmit signal energy, effect of noise can be at best minimized, not eliminated as in digital PCM transmission.

To obtain a better understanding for the bit error probability P_ϵ in (5.70), it can be lower bounded as follows. From there

$$P_\epsilon = Q(\alpha) = \int_\alpha^\infty \frac{1}{\sqrt{2\pi}} e^{-x^2/2}\, dx \tag{5.72}$$

where

$$\alpha = \left(\frac{1-\rho}{2}\right)\sqrt{SNR} \tag{5.73}$$

Pulse Code Modulation (PCM) Noise Analysis

Since

$$\alpha e^{-x^2/2} < x e^{-x^2/2} \quad , \quad x > \alpha \tag{5.74}$$

we get

$$\int_\alpha^\infty \alpha e^{-x^2/2}\, dx < \int_\alpha^\infty x e^{-x^2/2}\, dx = \int_{\alpha^2/2}^\infty e^{-y}\, dy = e^{-\alpha^2/2} \tag{5.75}$$

or

$$\int_\alpha^\infty \frac{1}{\sqrt{2\pi}} e^{-x^2/2}\, dx < \frac{1}{\alpha\sqrt{2\pi}} e^{-\alpha^2/2} \tag{5.76}$$

Substituting (5.76) into (5.72), we get

$$P_\epsilon \leq \frac{1}{(1-\rho)}\sqrt{\frac{2}{\pi \cdot SNR}} \cdot e^{-(1-\rho)^2\, SNR/8} \tag{5.77}$$

Pulse Code Modulation (PCM) Noise Analysis

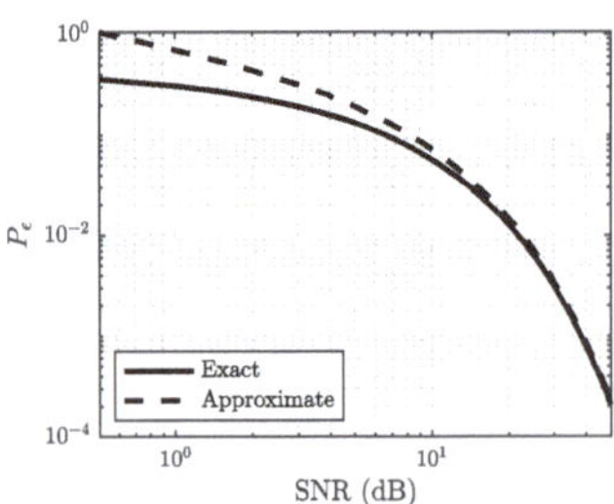

Figure 5.18: Bit error probability - exact and approximate

For BPSK, $\rho = -1$ and (5.77) reduces to

$$P_\epsilon < \frac{1}{\sqrt{2\pi \cdot SNR}} e^{-SNR/2} \tag{5.78}$$

From (5.77)-(5.78) and Fig. 5.18, the bit error probability P_ϵ decays exponentially with increase in SNR and that leads to practically error free digital data transmission and recovery.

Pulse Code Modulation (PCM) Noise Analysis

Table 5.3

SNR in dB	Bit error Prob. P_b	Prob. of error P_ϵ in bits/sec for a bit rate $f_0 = 1\,\text{MHz} = 10^6\,\text{Hz}$
11.4	10^{-4}	once every 0.01 sec.
13.6	10^{-6}	once every sec.
15.0	10^{-8}	once every 2 min.
16.0	10^{-10}	once every 2.5 hrs.
17.0	10^{-12}	once in 9 days

Thus for a bit rate of 1 MHz, channel noise in baseband PCM is virtually eliminated, provided the SNR in the received input signal is larger than 15dB. Using

$$SNR = E/\sigma^2$$

the input SNR can be easily adjusted by controlling the common pulse amplitude height E. However, the required bandwidth $f_0 = 2pf_s$, is $2p$ times the signal bandwidth f_s. Thus as in angle modulation, digital PCM trades off bandwidth for noise suppression as in frequency modulation.

Waveform Design and Modulation Schemes

BPSK refers to transmitting one bit (1 or 0) using two waveforms $s(t)$ and $-s(t)$ respectively. Multiple bits can be combined together to generate M-ary transmission schemes as described in (5.19).

QPSK (Quadrature Phase Shift Keying)
For example, in Quadrature Phase Shift Keying (QPSK), the waveform set $(\pm\cos\omega_0 t, \pm\sin\omega_0 t)$ is used to represent the data bit configurations in (5.19). This is shown in Fig. 5.19 along with 8-PSK representing 3 bits.

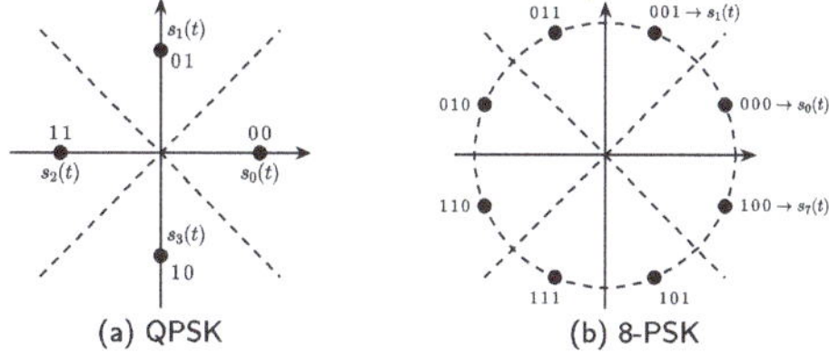

Figure 5.19: QPSK vs. 8-PSK and Gray Coding

Waveform Design and Modulation Schemes

The mapping of k information bit combinations to waveforms as in (5.19) can be done in any number of ways. A common method is to use Gray coding, where the bit combinations representing the adjacent signals differ by one binary digit as illustrated in Fig. 5.19.

M-ary PSK
The transmit waveforms in Fig. 5.19 representing M-ary PSK can be expressed as

$$s_k(t) = \sqrt{E}\cos(\omega_0 t + 2\pi/M) \quad , \quad 0 < t < T \; ; \; k = 0, 1, 2, \ldots, M-1 \tag{5.79}$$

M-ary Orthogonal Waveforms
If M-waveforms have the orthogonal property

$$s_1(t) \perp s_2(t) \perp \ldots \perp s_M(t) \tag{5.80}$$

Waveform Design and Modulation Schemes

i.e.

$$\int_0^T s_i(t)\, s_j(t) = 0 \quad , \quad i \neq j \tag{5.81}$$

then they represent an M-ary orthogonal scheme in Fig. 5.20

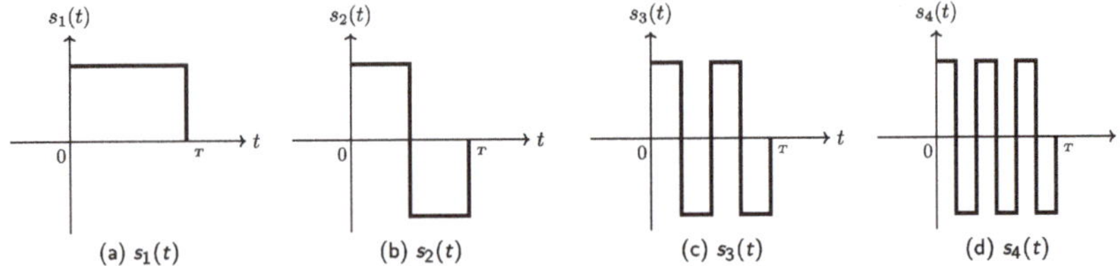

(a) $s_1(t)$ (b) $s_2(t)$ (c) $s_3(t)$ (d) $s_4(t)$

Figure 5.20: M-ary orthogonal Signaling

PAM (Pulse Amplitude Modulation)

If we choose to design waveforms by varying their amplitudes, that gives rise to Pulse Amplitude Modulation (PAM) as shown in Fig. 5.21.

Waveform Design and Modulation Schemes

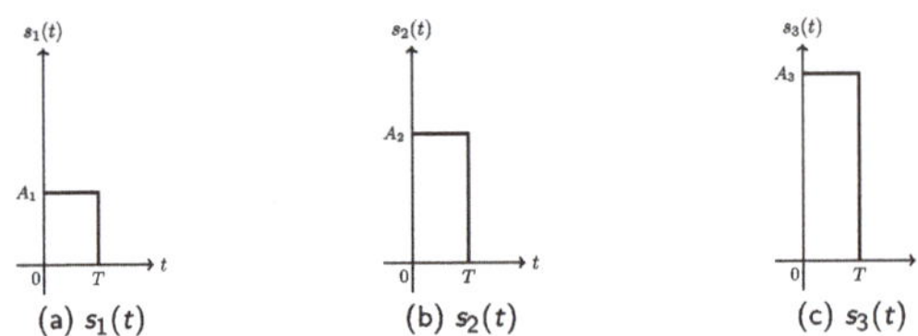

(a) $s_1(t)$ (b) $s_2(t)$ (c) $s_3(t)$

Figure 5.21: Pulse Amplitude Modulation

PWM (Pulse Width Modulation)

Pulses of different widths can be used to generate a set of waveforms as shown in Fig. 5.22.

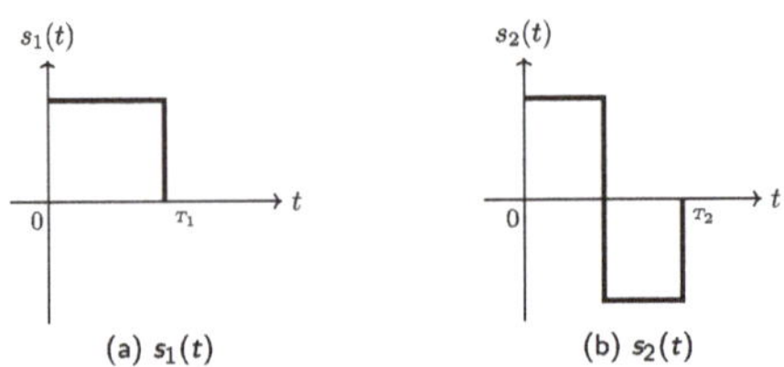

(a) $s_1(t)$ (b) $s_2(t)$

Figure 5.22: Pulse Width Modulation

Waveform Design and Modulation Schemes

QAM (Quadrature Amplitude Modulation)

By combining phase and amplitude for signal design, both PSK and PAM can be used to design Quadrature Amplitude Modulation (QAM) signal constellations as shown in Fig. 5.22.

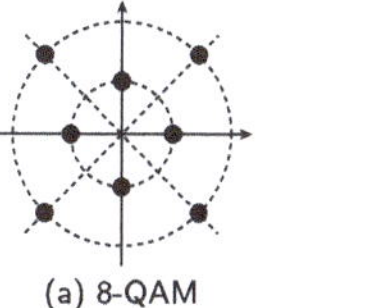

(a) 8-QAM (b) 12-QAM

Figure 5.23: PSK-PAM (QAM) signal constellations

In general, QAM waveforms have the form

$$s_k(t) = a_k \cos \omega_0 t + b_k \sin \omega_0 t \quad , \quad k = 1, 2, \ldots M \tag{5.82}$$

To understand the robustness of each of the above signal schemes and for their relative comparison, it is necessary to carry out the probability of error analysis for each such case. To illustrate this, probability of error analysis for M-ary orthogonal waveform scheme discussed above is carried out next.

Fourier Series Fourier Transforms Amplitude Modulation Angle Modulation Digital Modulation Techniques

Probability of Error Analysis for M-ary Orthogonal Signaling Scheme

In an M-ary orthogonal scheme, one of the M-signals is transmitted depending on the codeword present. Here $\{s_k(t)\}_{k=1}^{M}$ are such that

$$\int_0^T s_i(t) s_j(t)\, dt = 0 \quad , \quad i \neq j \tag{5.83}$$

and

$$\int_0^T {s_i}^2(t)\, dt = E \tag{5.84}$$

Thus under hypotheses H_k (that $s_k(t)$ is transmitted), the received signal is given by

$$r(t) = s_k(t) + n(t), \quad 0 < t < T, \quad H_k, \quad k = 1 \rightarrow n \tag{5.85}$$

where the input noise $n(t)$ is assumed to be zero mean white Gaussian noise with spectral density σ^2.

Fundamentals of Communication Theory U. Pillai & A. Patel

Probability of Error Analysis for M-ary Orthogonal Signaling Scheme

At the receiver, $r(t)$ is passed through a bank of matched filters $s_1(-t)$, $s_2(-t)$, $\ldots s_M(-t)$ as shown in Fig. 5.24

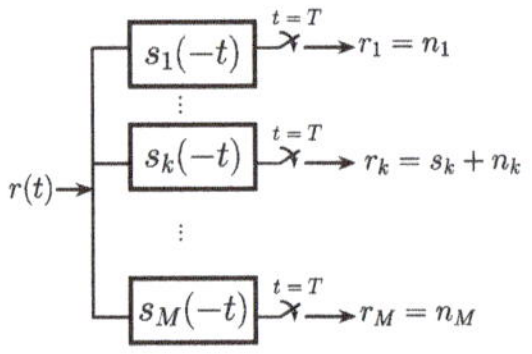

Figure 5.24: Receiver for M-ary orthogonal scheme. Output under hypothesis H_k.

The kth matched filter output at $t = T$ equals

$$r_k = r(t) \circledast s_k(-t) = \int_0^T r(t)s_k(t)\,dt$$
$$= \begin{cases} \int_0^T {s_k}^2(t)\,dt + \int_0^T s_k(t)n(t)\,dt = E + n_k \\ \int_0^T s_j(t)n(t)\,dt = n_j \quad , \quad j \neq k \end{cases} \tag{5.86}$$

Probability of Error Analysis for M-ary Orthogonal Signaling Scheme

Thus

$$r_k = \begin{cases} n_1 \\ n_2 \\ \vdots \\ E + n_k \quad , \quad \text{under } H_k \\ \vdots \\ n_M \end{cases} \tag{5.87}$$

i.e., if $s_k(t)$ was transmitted, then the kth receiver filter component in Fig. 5.24 will have a signal and noise component and all other outputs will have just noise components n_j, $j \neq k$.

Probability of Error Analysis for M-ary Orthogonal Signaling Scheme

All the noise components above are Gaussian and independent, since

$$\begin{aligned} E[n_i\, n_j] &= E\left[\int_0^T s_i(t_1)n(t_1)\,dt_1 \int_0^T s_j(t_2)n(t_2)\,dt_2\right] \\ &= \int_0^T\int_0^T s_i(t_1)s_j(t_2)\, \underbrace{E[n(t_1)n(t_2)]}_{\sigma^2\delta(t_1-t_2)}\, dt_1\, dt_2 \\ &= \sigma^2 \int_0^T s_i(t)s_j(t)\,dt = \begin{cases} \sigma^2 E & i=j \\ 0 & i \neq j \end{cases} \end{aligned} \tag{5.88}$$

The Gaussian property follows from the linearity in (5.86). Hence n_j in (5.87) are all independent zero mean Gaussian (Normal) random variables with common variance $\sigma^2 E$ as given by (5.88).

Probability of Error Analysis for M-ary Orthogonal Signaling Scheme

Thus (5.87) can be expressed as

$$r_i \sim \mathcal{N}(0, \sigma^2 E) \quad , \quad i \neq k \tag{5.89}$$

$$r_k \sim \mathcal{N}(E, \sigma^2 E) \quad , \quad \text{under } H_k \tag{5.90}$$

when $s_k(t)$ is transmitted (under H_k), the output $r_k > r_i$, $i = 1, 2, \ldots M (\neq k)$, and hence the probability of detection P_D under H_k is given by the conditional probability

$$\begin{aligned} P(D|H_k) &= P(r_1 < r_k,\, r_2 < r_k \ldots r_M < r_k \,|\, H_k) \\ &= \int_{-\infty}^{\infty} P(r_1 < r_k,\, r_2 < r_k \ldots r_M < r_k \,|\, r_k) f_{r_k}(r_k)\,dr_k \\ &= \int_{-\infty}^{\infty} \underbrace{P(r_1 < r_k \,|\, r_k) \ldots P(r_M < r_k \,|\, r_k)}_{M-1 \text{ factors}} f_{r_k}(r_k)\,dr_k \end{aligned} \tag{5.91}$$

where the last step follows from the independence of $r_1, r_2 \ldots r_M$ given r_k.

Probability of Error Analysis for M-ary Orthogonal Signaling Scheme

Using (5.89)

$$P(r_i < r_k \mid r_k) = \int_{-\infty}^{r_k} \frac{1}{\sqrt{2\pi\sigma^2 E}} e^{-x^2/2\sigma^2 E}\, dx = \int_{-\infty}^{r_k/\sigma\sqrt{E}} \frac{1}{\sqrt{2\pi}} e^{-y^2/2}\, dy$$
$$= 1 - Q\left(\frac{r_k}{\sigma\sqrt{E}}\right) \tag{5.92}$$

where we made use of (5.66) and $y = \frac{x}{\sigma\sqrt{E}}$. Substituting (5.92) into (5.91), we get

$$\begin{aligned} P(D \mid H_k) &= \int_{-\infty}^{\infty} \left[1 - Q\left(\frac{r}{\sigma\sqrt{E}}\right)\right]^{M-1} f_{r_k}(r)\, dr \\ &= \int_{-\infty}^{\infty} \left[1 - Q\left(\frac{r}{\sigma\sqrt{E}}\right)\right]^{M-1} \frac{1}{\sqrt{2\pi\sigma^2 E}} e^{\frac{-(r-E)^2}{2\sigma^2 E}}\, dr \end{aligned} \tag{5.93}$$

Probability of Error Analysis for M-ary Orthogonal Signaling Scheme

Let $\frac{r-E}{\sigma\sqrt{E}} = x$ so that $dr = \sigma\sqrt{E}$, $\frac{r}{\sigma\sqrt{E}} = x + \sqrt{\frac{E}{\sigma^2}} = x + \sqrt{SNR}$; so that (5.93) simplifies to

$$P(D \mid H_k) = \int_{-\infty}^{\infty} \left[1 - Q\left(x + \sqrt{SNR}\right)\right]^{M-1} \frac{1}{\sqrt{2\pi}} e^{-x^2/2}\, dx \tag{5.94}$$

and finally the probability of detection P_D equals

$$P_D = \sum_{k=1}^{M} P(D|H_k)P(H_k) = \int_{-\infty}^{\infty} \left[1 - Q\left(x + \sqrt{SNR}\right)\right]^{M-1} \frac{1}{\sqrt{2\pi}} e^{-x^2/2}\, dx \tag{5.95}$$

and

$$P_\epsilon = 1 - P_D \tag{5.96}$$

represents the common symbol error probability.

Probability of Error Analysis for M-ary Orthogonal Signaling Scheme

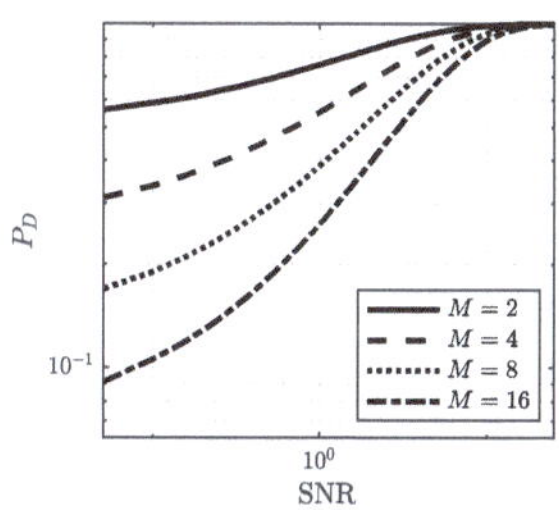

Figure 5.25: Probability of detection vs. number of signals

If the M signals are used to represent k bits then $M = 2^k$ and $E = kE_b$, where E_b represents the bit energy, then the corresponding bit error probability $P_b \simeq P_\epsilon$, is given by (5.96) with symbol energy E replaced by kE_b.

$$P_b \simeq P_\epsilon = 1 - \frac{1}{\sqrt{2\pi}} \int_{-\infty}^{\infty} \left[1 - Q\left(x + \sqrt{\frac{E_b \cdot \ln M}{\sigma^2}} \right) \right]^{M-1} e^{-x^2/2}\, dx \tag{5.97}$$

Fourier Series Fourier Transforms Amplitude Modulation Angle Modulation Digital Modulation Techniques

References

[1] J. G. Proakis and M. Salehi, *Digital communications.* McGraw-Hill, New York, 2008.

[2] J. G. Proakis and M. Salehi, *Communication Systems Engineering*, 2nd ed., Prentice Hall, New Jersey, 2002.

[3] B. P. Lathi and Z. Ding, *Modern Digital and Analog Communication Systems.* Oxford University Press, United Kingdom, 2009.

[4] G. P. Tolstov, *Fourier Series.* Dover Publications, New York, 1976.

[5] I. N. Sneddon, *Fourier Transforms.* Dover Publications, New York, 1995.

[6] G. N. Watson, *A Treatise on the Theory of Bessel Functions.* Cambridge University Press, United Kingdom, 1996.

[7] P. Dyke, *An Introduction to Laplace Transforms and Fourier Series.* Springer London, United Kingdom, 2014.

[8] A. Papoulis and S. U. Pillai, *Probability, Random Variables, and Stochastic Processes*, 4th ed., McGraw-Hill, New York, 2002.

Fundamentals of Communication Theory U. Pillai & A. Patel

Zeitfracht Medien GmbH
Ferdinand-Jühlke-Straße 7
99095 Erfurt, Deutschland
produktsicherheit@kolibri360.de